Advances in Industrial Control

Other titles published in this series:

Matjaž Colnarič • Domen Verber
Wolfgang A. Halang

Distributed Embedded Control Systems

Improving Dependability with Coherent Design

 Springer

Prof. Dr. Matjaž Colnarič
University of Maribor
Faculty of Electrical Engineering and
   Computer Science
2000 Maribor
Slovenia

Prof. Dr. Dr. Wolfgang A. Halang
Faculty of Electrical and Computer
   Engineering
FernUniversität in Hagen
58084 Hagen
Germany

Dr. Domen Verber
University of Maribor
Faculty of Electrical Engineering and
   Computer Science
2000 Maribor
Slovenia

ISBN 978-1-84800-051-3   e-ISBN 978-1-84800-052-0

DOI 10.1007/978-1-84800-052-0

Advances in Industrial Control series ISSN 1430-9491

British Library Cataloguing in Publication Data
A catalogue record for this book is available from the British Library

Library of Congress Control Number: 2007939804

Cover design: eStudio Calamar S.L., Girona, Spain

Printed on acid-free paper

9 8 7 6 5 4 3 2 1

springer.com

Advances in Industrial Control

Professor Emeritus O.P. Malik
Department of Electrical and Computer Engineering
University of Calgary
2500, University Drive, NW
Calgary
Alberta
T2N 1N4
Canada

Professor K.-F. Man
Electronic Engineering Department
City University of Hong Kong
Tat Chee Avenue
Kowloon
Hong Kong

Professor G. Olsson
Department of Industrial Electrical Engineering and Automation
Lund Institute of Technology
Box 118
S-221 00 Lund
Sweden

Professor A. Ray
Pennsylvania State University
Department of Mechanical Engineering
0329 Reber Building
University Park
PA 16802
USA

Professor D.E. Seborg
Chemical Engineering
3335 Engineering II
University of California Santa Barbara
Santa Barbara
CA 93106
USA

Doctor K.K. Tan
Department of Electrical Engineering
National University of Singapore
4 Engineering Drive 3
Singapore 117576

Professor Ikuo Yamamoto
The University of Kitakyushu
Department of Mechanical Systems and Environmental Engineering
Faculty of Environmental Engineering
1-1, Hibikino,Wakamatsu-ku, Kitakyushu, Fukuoka, 808-0135
Japan

We wish to dedicate this book to our families in gratitude of their support during the last fifteen years of work on this research.

Series Editors' Foreword

The series *Advances in Industrial Control* aims to report and encourage technology transfer in control engineering. The rapid development of control technology has an impact on all areas of the control discipline. New theory, new controllers, actuators, sensors, new industrial processes, computer methods, new applications, new philosophies..., new challenges. Much of this development work resides in industrial reports, feasibility study papers and the reports of advanced collaborative projects. The series offers an opportunity for researchers to present an extended exposition of such new work in all aspects of industrial control for wider and rapid dissemination.

Embedded systems are computer systems designed to execute a specific task or group of tasks. In the parlance of the subject, an embedded system has dedicated functionality. Looking at the hardware of an embedded system one would expect to find a small unified module involving a microprocessor, a Random Access Memory unit, some task-specific hardware units and even mechanical parts that would not be found in a more general computer system. The objective of a dedicated functionality means that the design engineer can optimise hardware and software components to achieve the required functionality in the smallest possible size, with good operational efficiency and at reduced cost. If the application is to be mass-produced, economies of scale often play an important role in reducing the costs involved.

From an applications viewpoint there are two aspects to embedded systems:

- low-level aspects; these involve microprocessor-based, real-time computer system design and optimisation. To achieve the dedicated-functional objectives of the embedded system, the internal tasks are performed sequentially and in a temporally feasible manner;
- high-level aspects; the applications for embedded systems can be simple using only one or two system modules to achieve a few high-level tasks as might be needed in a central-heating system controller or digital camera. In more complex applications, there may be dozens of embedded systems

working in concert, organised in a hierarchical multi-level network communicating low-level sensory information (collected by dedicated embedded system modules) to high-level processors that will direct actuators to control a complex process. Typical applications are holistic automobile control systems or the control of a highly dynamical industrial process like a steel mill or an avionics system used in aircraft flight control.

Clearly, embedded systems are extremely important in industrial control system implementation, providing, as they do, the hardware and software infrastructure for each application whether simple or complex. Professors Matjaž Colnarič, Domen Verber and Wolfgang Halang have devoted many years' study to the design of the architectures for embedded system modules. They have been supported in their research by European Union funding mechanisms for the EU has been very concerned to promote expertise in embedded system technologies. This *Advances in Industrial Control* monograph reports their important research. They have divided their monograph into two parts; the first part is devoted to concepts and guidelines and the second is concerned with implementation. The monograph will be of considerable interest to the wide readership of academic and industrial practitioners in control engineering.

Industrial Control Centre   *M.J. Grimble*
Glasgow   *M.A. Johnson*
Scotland, UK

Preface

This book is a result of 15 years of relatively intensive co-operation. All this time, we have been dealing with proper design of safety-related embedded systems, considering many domains in a holistic way. We started with concepts, and have proposed hypothetical hardware and system architectures, together with programming means. We have also implemented a couple of prototypes. Now, as our common research has reached a stage that many of the pertinent domains have been dealt with to a reasonable extent, we thought it was time to publish our results.

To promote adequate and consistent design of embedded systems with dependability requirements, this book is primarily dedicated to practitioners and specialists, as well as to students in computer, electrical and automation engineering. In order to provide information useful to them, for each topic we present both basic considerations and examples of use and/or implementation. In this sense, this book's role is at least twofold. First, it is intended to help designers of control applications to select and design appropriate solutions and, second, to provide some ideas and case studies from on-going research into the topics, related to the further elaboration of hardware and software solutions to be employed in real-time control systems.

The book is structured in two parts. In Part I, long established concepts are presented, which we find to be most important and suitable for the implementation of embedded control systems. This part could also serve as a textbook for courses covering embedded real-time systems. In Part II, the approaches and solutions to implement prototypes of embedded systems are detailed, which were jointly devised by the authors. Some of them also originate from the 5th Framework EU project IFATIS, which dealt with reconfiguration as a means to achieve fault tolerance, and which was successfully concluded in March 2005.

What we offer in this book, and particularly in Part II, is not to be considered as the only solutions possible, probably not even the most adequate or applicable ones, but as possible solutions coherent with commonly accepted guidelines. Their feasibility was shown by prototype implementations in our laboratory, and by utilisation in process control projects.

This book could not have been realised without substantial contributions from a number of persons. Most of the practical work described has been carried out in the Real-Time Systems Laboratory of the Faculty of Electrical Engineering and Computer Science at the University in Maribor, Slovenia. In this framework, three doctoral theses have successfully been concluded, jointly supervised by the authors, and a number of journal and conference papers has been co-authored. Thus, we should like to express our sincere appreciation to the members of the Real-Time Systems Laboratory who participated in the research on embedded systems' design, *viz.*, to Dr. Roman Gumzej, Dr. Matej Šprogar, Rok Ostrovršnik, Stanislav Moraus, and Bojan Hadjar. In particular, Dr. Matej Šprogar worked on time-triggered communication, Dr. Roman Gumzej elaborated certain issues in hardware/software co-design and specification of embedded real-time systems, and Rok Ostrovršnik implemented the system for designing embedded applications in MATLAB®/Simulink®. Stanislav Moraus and Bojan Hadjar worked on the technical implementation of the prototypes. Finally, Dr. Šprogar thoroughly proof-read the texts for technical errors and consistency. A special chapter on implementation of embedded systems from his doctoral thesis, jointly supervised at Fernuniversität in Hagen, and some other parts (specifically, history of safety standards and comparison of rate-monotonic and earliest-deadline-first scheduling) have been prepared by Dr.-Ing. Martin Skambraks. Last but not least, our thanks go to Springer-Verlag's assistant editor Oliver Jackson for his encouragement, support and, most of all, his patience.

Matjaž Colnarič
Domen Verber
Wolfgang A. Halang

Maribor, Hagen,
September 2007

Contents

Part II Implementation

Part I

Concepts

1

Real-time Characteristics and Safety of Embedded Systems

1.1 Introduction

What an embedded system is, is not exactly defined. In general, this term is understood to mean a special-purpose computer system designed to control or support the operation of a larger technical system (termed the embedding system) usually having mechanical components and in which the embedded system is encapsulated. Unlike a general-purpose computer, it only performs a few specific, more or less complex pre-defined tasks. It is expected to function without human interaction and, therefore, it usually has sensors and actuators, but not peripheral interfaces like keyboards or monitors, except if the latter are required to operate the embedding system. Often, it functions under real-time constraints, what means that service requests must be handled within pre-defined time intervals.

Embedded systems are composed of hardware and corresponding software parts. The complexity of the hardware ranges from very simple programmable chips (like field programmable gate arrays or FPGAs) over single microcontroller boards to complex distributed computer systems. Usually the software is stored in ROMs, as embedded systems seldom have mass storage facilities. Peripheral interfaces communicate with the process environments, and usually include digital and analogue inputs and outputs to connect with sensors and actuators.

In simpler cases, the software consists of a single program running in a loop, which is started on power-on, and which responds to certain events in the environment. In more complex cases, operating systems are employed, providing features like multitasking, different scheduling policies, synchronisation, resource management and others, to be dealt with later in this book.

The trend towards distributed architectures away from centralised ones assures modularity for structured design, better distribution of processing power, robustness, fault tolerance, and other advantages.

There are almost no areas of modern technology which could do without embedded systems. They appear in all areas of industrial applications and process control, in cars, in home appliances, entertainment electronics, cellular phones, video and photo cameras, and many more places. We even have them implanted, or wear them in our garments, shoes, or eye glasses. Their major spread has occurred particularly in the last decade. They pervade areas where they were only recently not considered. As they are becoming ubiquitous, we gradually do not notice them any more.

Contemporary cars, for example, contain dozens of embedded computers connected *via* hierarchically organised multi-level networks to communicate low-level sensory information or inter-processor messages, and to provide higher application-level interconnection of multimedia appliances, navigation systems, *etc.* The driver is not aware of the computers involved, but is merely utilising the new functionality. Within such automotive systems, there are also safety-critical components being prepared to deal in the near future with functions like drive-, brake-, or steer-by-wire. Should such a system fail, the consequences are much different than for, *e.g.*, Anti-Blocking Systems, whose function simply ceases in the case of a failure without putting users in immediate danger. With to the failure of an x-by-wire facility, however, drivers would not even be in a position to stop their cars safely.

Such considerations brought into light another aspect that has not been observed before. Since in the past embedded systems were considered to be sensitive high-technology elements, they were observed with a certain amount of precautious scepticism and doubt in their proper functioning. Special care was taken in their implementation, and they were not employed in the most safety-critical environments, like control units of nuclear power plants.

As a consequence of the increasing complexity of control algorithms in such applications, for better flexibility, and for economic reasons, however, and of getting used to their application in other areas, embedded systems have also found their way into more safety-critical areas where the integrity of the systems substantially depends on them. Any failures could have severe consequences: they may result in massive material losses or endanger human safety. Often the implementation of embedded systems is inadequate with regard to the means and/or the methods employed. Therefore, it is the prime goal of this book to point out what should be considered in the various design domains of embedded systems. A number of long existing guidelines, methods, and technologies of proper design will be mentioned and some elaborated in more detail.

By definition, embedded systems operate in the real-time domain, which means that their temporal behaviour is — at least — equally as important as their functional behaviour. This fact is often not considered seriously enough. There are a number of misconceptions that have been identified in an early paper by Stankovic [104]; some characteristic and still partially valid ones will be elaborated later in this chapter.

While verifying embedded systems' conformance to functional specifications is well established, temporal circumstances are seldom consistently verified. The methods and techniques employed are predominantly based on testing, and the quality achieved mainly depends on the experience and intuition of the designers. It is almost never proven at design time that such a system will meet its temporal requirements in every situation that it may encounter.

Unfortunately, this situation was identified more than 20 years ago, when the basic principles of the real-time research domain was already well organised. Although adequate partial solutions were known for a number of years, in practice embedded systems design did not progress essentially during this time. Therefore, in this work we present certain contributions to several critical areas of control systems design in a holistic manner, with the aim to improve both functional and temporal correctness. The implementation of long established, but often neglected, viable solutions will be shown with examples, rather than devising new methods and techniques. As verification of functional correctness is more established than that of temporal correctness, although equally important, special emphasis will be given to the latter.

While adequate verification of temporal and functional behaviour is important for high quality design of embedded systems, it cannot be taken as a sufficient basis to improve their dependability. It is necessary to consider the principles of fault management and safety measures for such systems in the early design phases, which means that common commercial off-the-shelf control computers are usually unsuitable for safety-critical applications.

In the late 1980s, the International Electrotechnical Commission (IEC) started the standardisation of safety issues in computer control [58]. It identified four Safety Integrity Levels (*SIL*), with SIL 4 being the most critical one (more details follow in Section 1.3.2). This book, however, is concerned with applications falling into the least demanding first level SIL 1, which allows the use of computer control systems based on generic microprocessors. It is desirable that such systems should formally be proven correct or even be safety-licensed. Owing to the complexity of software-based computer control systems, however, this is very difficult if not impossible to achieve.

1.2 Real-time Systems and their Properties

Let us start with some examples that demonstrate what real-time behaviour of a system actually is. A very good, but unexpected, example of proper and problem-oriented temporal behaviour, dynamic handling of priorities, synchronisation, adaptive scheduling and much more is the daily work of a housekeeper and parent, whose tasks are to care for children, to do a lot of housework and shopping, and to cook for the family. Apart from that, the housekeeper also receives telephone calls and visitors. Some of these tasks are known in advance and can be statically planned (scheduled), like sending children to school, doing laundry, cooking lunch, or shopping. On the other hand, there are others that happen sporadically, like a visit of a postman, telephone calls, or other events, that cannot be planned in advance. The reactions to them must be scheduled dynamically, *i.e.*, current plans must be adapted when such events occur.

For statically scheduled tasks, often a chain of activities must be properly carried through. For instance, to send the children to the school bus, they must be woken on time, they must use the bathroom along with other family members, enough time must be allowed for breakfast which is prepared in parallel with the children being in the bathroom and getting dressed. The deadline is quite firm, namely, the departure of the school bus. In the planning, enough time must be allocated for all these activities. It is not a good idea, however, to allow for too much slack, since the children should not have to get up much earlier than necessary, thus losing sleep in the morning.

After sending the children to school, there are further tasks to be taken care of. Housekeeping, laundry, cooking and shopping are carried out in an interleaved manner and partly in parallel. Some of these tasks have more or less strict deadlines (*e.g.*, lunch should be ready for the children coming in from school). The deadlines can be set according to the time of the day (or the clock) or relative to the flow of other events. If the housekeeper is cooking eggs or boiling milk, the time until they will be ready is known in advance. If a sporadic event like a telephone call or postman's visit occurs during that time, the housekeeper must decide whether to accept it or not. If the event is urgent, it may be decided to re-schedule the procedure and interrupt cooking until the event is taken care of. Needless to say, that there are events with high and absolute priorities that will be handled regardless of other consequences; if, for example, a child is approaching a hot electric iron, then the housekeeper will interrupt any other activity whatsoever, even at the cost of milk boiling over.

Knowing his or her resources well, the housekeeper behaves very rationally. If, for instance, food provisions are kept in the same room where the laundry is done, the housekeeper will collect the vegetables needed for cooking when going there to start the washing machine, although they will not be needed until a later stage in the course of the housework planned, *e.g.*, after having made the beds.

1.2.1 Definitions, Classification and Properties

Following the pattern of the above example, in technical control systems there is usually a process that needs to be carried through. A *process* is the totality of activities in a system which influence each other and by which material, energy, or information is transformed, transported, or stored [28]. Specifically, a technical process is a process dealing with technical means. The basic element of a process is the *task*. It represents the elementary and atomic entity of parallel execution. The task concept is fundamental for asynchronous programming. It is concerned with the execution of a program in a computing system during the lifetime of a process.

Considering the housewife example again, it is interesting that very complex sequences of tasks are quite normal for ordinary people like in the housekeeper example, and are carried out just with common sense. In the so-called "high technology" world of computers, however, people are reluctant to consider similar problems that way. Instead, sophisticated methods and procedures are devised to match obsolete approaches that were used in the past due to under-development, such as static priority scheduling.

Control systems should be considered in terms of tasks with their inherent natural properties. Each one's urgency is expressed by its deadline and not by artificially assigned priorities. This concept matches the natural behaviour of the housewife, as it is her goal to *perform her tasks in such a sequence and schedule that each tasks will be completed before its deadline*. This natural perception of tasks, priorities and deadlines is the essence of real-time behaviour:

> *In the real-time operating mode of a computer system the programs for the processing of data arriving from the outside are permanently ready, so that their results will be available within predetermined periods of time [27].*

Let us now consider two further examples that will lead us to a classification of real-time systems.

In preparation for a journey, we visit a travel agent to book a flight and buy tickets. The agent's job is to see which flights are available, to check the prices, and to make a reservation. If the service is busy, or there are any other unfortunate circumstances, this can take some time, or could even not be completed during our margin of patience. In the latter case, the agent could not fulfill the job, and we did not get our tickets. The deadline that has not been met was not very firmly set; it depended on a number of circumstances, *e.g.*, we were in a hurry or in a bad mood. Also, the longer we had to wait, the higher the probability that we would go to another agent next time.

When we go to the airport after the booking, the deadlines are set differently: if we are for some reason late and arrive after the door is closed (that deadline was known to us in advance), we have failed. It does not matter if we were late only by a few seconds or an hour. It does not even matter if we made any other functional mistake, for example went to wrong airport: it is the same if the failure to board was due to a functional or temporal error.

Considering the two examples above, we can classify the real-time systems into two general categories: systems with *hard and soft real-time behaviour*. Their main difference lies in the cost or penalty for missing their deadlines (see Figure 1.1). In the case of soft real-time systems, like in our example of flight ticketing, after a certain deadline the costs or penalty (customer dissatisfaction and, consequently, possibility of losing the customer) begin to rise. After a certain time, the action can be considered to have failed.

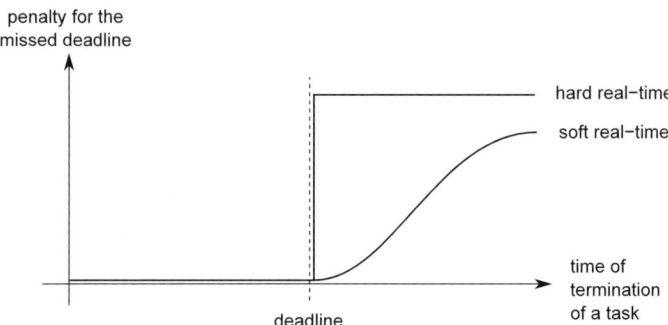

Fig. 1.1. Soft *vs* hard real-time temporal behavioural

In the case of hard real-time systems, as in our second example of missing a flight, the action has failed immediately after the deadline is missed. The cost or penalty function exhibits a jump to a certain high value indicating total failure, which may be high material costs or even endangering of environmental or human safety. Hence, hard real-time systems are those for which it holds that:

> *although functionally correct, the results produced after a certain predefined deadline are incorrect and, thus, useless.*

A task's failure to perform its function in due time may have different consequences. According to them, the hard- or soft real-time task may or may not be *mission-critical*, *i.e.*, the main objective of the entire application could not be met. Sometimes, however, a failure of a task can be tolerated, *e.g.*, when as a consequence only the performance is reduced to a certain extent. For instance, MPEG video-decoders in multimedia applications operate in the hard real-time mode: if a frame could not be decoded and composed before it would have to be put on screen, which is a hard deadline, the task failed as the frame is missing. The consequence would be flickering, which can be tolerated if it does not happen often — thus, it is not mission-critical.

On the other hand, soft real-time systems can be *safety-critical*. As an example, let us consider a diagnostics system whose goal is to report a situation of alert. Since human reaction times are relatively long and variable, it is not sensible to require the system's reaction to be within a precisely defined timeframe. However, the action's urgency increases with delay. The soft real-time deadline has a very positive side effect, namely, it allows other tasks more time to deal with the situation causing the alert and possibly to solve it.

Figure 1.1 depicts, and the definitions describe, two extreme cases of hard and soft real-time behaviour. In reality, however, the boundaries are often not so strict. Moreover, beside cost, benefit functions may also be considered, and different curves can be drawn [97]. Jensen describes the problem colourfully:

"They (the real-time research community) have consensus on a precise technical (and correct) definition of "hard real-time," but left "soft real-time" to be tautologically defined as "not hard" — that is accurate and precise, but no more useful than dichotomising all colours into "black" and "not black" [67].

Together with Gouda and others [44] he has further elaborated the issue with "Time/Utility Functions" based on earliness, tardiness and lateness.

From the above we can conclude that *predictability of temporal behaviour* is the ultimate property of real-time systems. The necessary condition is determinism of temporal behaviour of the (sub-) systems. Strict and realistic predictability, however, is very difficult to achieve — practically impossible regarding the hardware and system architectures as employed in state-of-the-art embedded control systems. Hence, a much more pragmatic approach is needed.

In [105], Stankovic and Ramamritham elaborate two different approaches to predictability: the layer-by-layer (microscopic) and the top-layer (macroscopic) approach. The former stands for low-level predictability which is derived hierarchically: a layer in the design of a real-time system (processor, system architecture, scheduling, operating system, language, application) can only be predictable if all underlying layers are predictable. This type of predictability is necessary for low-level critical parts of real-time systems, and it should be provable.

For the higher layers (real-time databases, artificial intelligence, and other complex controls) microscopic predictability cannot be achieved. In these cases it is important that best effort is to be devoted, and that temporal behaviour is observed. The goal is to meet the deadlines in most cases. However, since it was not possible to prove that they are met in all cases, provisions should be made for the rare occasions of missed deadlines. Fault tolerance means should be implemented to resolve this situation. These must be simple and, thus, provably predictable in the microscopic sense.

The history of systematic research into real-time systems goes back at least to the 1970s. Although many solutions to the essential questions have been found very early, there are still many misconceptions that characterise this domain. In 1988, Stankovic collected many of them [104]. He found that one of the most characteristic misconceptions in the domain of hard real-time systems is that real-time computing is often considered as fast computing; probably to a lesser extent, this misconception is still alive. It is obvious from the above-mentioned facts that computer speed itself cannot guarantee that specified timing requirements will be met. Instead, predictability of temporal behaviour has been recognised as the ultimate objective. Being able to assure that a process will be serviced within a predefined timeframe is of utmost importance. Thus

A computer system can be used in real-time operating mode if it is possible to prove at design time that in all cases all requests will be served within predefined timeframes.

Beside timeliness, which is ensured by predictability, another requirement real-time systems should fulfill is *simultaneity*. This property is more severe, especially in multitasking and multiprocessor environments. It involves the demand that the execution behaviour of a process should be timely even in the presence of other parallel processes, whose number and behaviour are not known at design time and with whom it will share resources. It is not always possible to prove this property, but it should be considered and best efforts made.

Finally, real-time systems are inherently safety-related. For that reason, real-time systems should be dependable which, beside the properties of functional and temporal correctness, also includes robustness and permanent readiness. This property renders them particularly hard to design. The safety issues will be elaborated later in this chapter.

1.2.2 Problems in Adequate Implementation of Embedded Applications and General Guidelines

Although guidelines for proper design and implementation of embedded control systems operating in real-time environments have been known for a long time, in practice *ad hoc* approaches still prevail to a large extent. There are some major causes for this phenomenon:

- The basic problem seems to be the mismatch between the design objectives of generic universal computing and embedded control systems. It is reasonable to employ various low-level (caching, pipelining, *etc.*) and high-level measures (dynamic structures, objects, *etc.*) to achieve the best possible average performance with universal computers. Often, these measures are based on improvement of statistical properties and are, thus, in contradiction to the ultimate requirement of real-time systems, *viz.*, temporal determinism and predictability. There are no modern and powerful processors with easily predictable behaviour, nor compilers for languages that would prevent us from writing software with non-predictable run times. Practically all dynamic and "virtual" features aiming to enhance the *average* performance of non-real-time systems are, therefore, considered harmful. Inappropriate categories and optimality criteria widely employed in systems design are probabilistic and statistical terms, fairness in task processing, and minimisation of average reaction time. In contrast to this, the view adequate for real-time systems can be characterised by observation of hard timing constraints and worst cases, prevention of deadlocks, prevention of features taking arbitrarily long to execute, static analysis, and recognition of the constraints imposed by the real, *i.e.*, physical, world.

- The costs of consistently designed real-time embedded applications are much higher than conventional software. Timing circumstances need to be considered in all design stages, from specification to maintenance. Especially the verification and validation phases, when performed properly, are much more demanding and costly than in conventional computing.
- Designers of embedded systems are often reluctant to observe guidelines for proper design. Often overloaded, they tend to develop their applications in the usual way that was more or less appropriate in previous projects, but may fail in a critical situation. Owing to lack of time, knowledge, and will, they are not prepared to do the hard, annoying and time-consuming work of proving their designs' functional and temporal correctness.

The notion of time has long been ignored as a category in computer science. It is suggested in a natural way by the flow of occurrences in the world surrounding us. As the fourth dimension of our (Euclidean) space of experience, time is already a model defined by *law* and technically represented by Universal Time Co-ordinated (UTC). Time is an absolute measure and a practical tool allowing us to plan processes and future events easily and predictably with their mutual interactions requiring no further synchronisation. This is contrasted by the conceptual primitivity of computing, whose central notion algorithm is time-independent. Here, time is reduced to predecessor-successor relations, and is abstracted away even in parallel systems. No absolute time specifications are possible, the timing of actions is left implicit in real-time systems, and there are no time-based synchronisation schemes. As a result, the poor state of the "art" is characterised by computers using interval timers and software clocks with low (and in operation decreasing) accuracy, which are much more primitive than wrist watches. Moreover, meeting temporal conditions cannot be guaranteed, timer interrupts may be lost, every interrupt causes overhead, and clock synchronisation in distributed systems is still assumed to be a serious problem, although radio receivers for official date and time signals, as already available for 100 years and widely used for many purposes, providing the precise and worldwide only legal time UTC could easily and cheaply be incorporated in any node.

The core problem of contemporary information technology, however, is complexity, which is particularly severe in embedded systems design. It can be observed that people tend to use sophisticated and complicated measures and approaches when they feel that they need to provide good real-time solutions for demanding and critical applications. It is, however, much more appropriate to find simple solutions, which are transparent and understandable and, thus, safer. Simplicity is a means to realise dependability, which is the fundamental requirement of safety-related systems. (Easy) understandability is the most important precondition to prove the correctness of real-time systems, since safety-licensing (verification) is a social process with a legal quality.

There is a large number of examples for extensive complexity, or better, *"artificial complicatedness"*. Thus, for instance, the standard document

DIN 19245 of the fieldbus system Profibus consists of 750 pages, and a tele-phone exchange, which burned down in Reutlingen, had an installed software base of 12 million lines of code. On the other hand, a good example of success-fully employing simple means in a high-technology environment is the general purpose computer used for the Space Shuttle's main control functions. It is based on five redundant IBM AP-101S computer systems whose development started in 1972; the last revision is from 1984, and it was deployed in 1991. They come out with 256k of 32 bit words of storage, and were programmed in the high-level assembly language HAL. Simplicity and stability of the design ensure the application's high integrity.

A serious problem in the design of safety-critical embedded systems is dependability of software:

We are now faced with a society in which the amount of software is doubling about every 18 months in consumer electronic devices, and in which software defect density is more or less unchanged in the last 20 years.

In spite of this, we persist in the delusion that we can write software sufficiently well to justify its inclusion at the highest levels of safety criticality.

Considering, for instance, the mean time between failure of a typical modern disk of around 500,000 h, the widening gulf between software quality and hardware quality becomes even more emphatic, to the point that the common procedure in safety critical systems of triplicating the same incredibly reliable hardware system and running the same much less reliable software in each channel seems questionable to say the least [52].

Software must be valid and correct, which means that it must fulfil its problem specification. For the validity of specifications there is no more au-thority of control — except the developers' wishes, or more or less vaguely formulated requests. In principle, automatic verification is possible. Valida-tion, on the other hand, is inherently hard, because it involves the human element to a great extent.

Software always contains design errors and, thus, needs correctness proofs, as tests cannot show the absence of errors. Safety-licensing of systems, whose behaviour is largely program-controlled, is still an unsolved problem, whose severity is increased by the legal requirement that verification must be based on object code. The still too big semantic gap between specifications on one hand and the too low a level programming constructs available on the other can be coped with by *the-other-way-around approach, viz.,* to select program-ming and verification methods of the utmost simplicity and, hence, highest trustworthiness, and to custom-tailor execution platforms for them.

Descartes (1641) pointed out the very nature of verification, which is nei-ther a scientific nor a technical, but a *cognitive process*:

Verum est quod valde clare et distincte percipio.[1]

Verification is also a *social process*, since mathematical proofs rely on consensus between the members of the mathematical community. To verify safety-related computerised systems, this consensus ought to be as wide as possible. Furthermore, verification has a legal quality as well, in particular for embedded systems whose malfunctioning can result in liability suits. Simplicity can be used as the fundamental design principle to fight complexity and to create confidence. Based on simplicity, easy understandability of software verification methods — preferably also for non-experts — is the most important precondition to prove software correctness.

Design-integrated verification with the quality of mathematical rigour and oriented at the *comprehension capabilities of non-experts* ought to replace testing to facilitate safety-licensing. It should be characterised by simple, inherently safe programming — better specification, re-use of already licensed application-oriented modules, graphics instead of text, and rigorous — but not necessarily formal — verification methods understandable by non-experts such as judges. The more *safety-critical* a function is, the more *simple* the related software and its verification ought to be.

> *Simple solutions are the most difficult ones: they require high innovation and complete intellectual penetration of issues.*
>
> *Progress is the road from the primitive* via *the complicated to the simple.*
>
> (Biedenkopf, 1994)

1.3 Safety of Embedded Computer Control Systems

> *To err is human, but to really foul things up requires a computer.*
> (Farmers' Almanac, 1978)

As society increasingly depends on computerised systems for control and automation functions in safety-critical applications and, for economical reasons, it is desirable to replace hardwired logic by programmable electronic systems in safety-related automation, there is a big demand for highly dependable programmable electronic systems for safety-critical embedded control and regulation applications. This domain forms a relatively new field, which still lacks its scientific foundations. Its significance arises from the growing awareness for safety in our society on the one hand, and from the technological trend towards more flexible, *i.e.*, program controlled, automation devices on the other hand. It is the aim to reach the state that computer-based systems can be constructed with a sufficient degree of confidence in their dependability.

---

[1] That which I perceive very clearly and distinctly is true.

Let us start with an example of a fault-tolerant design. In the Airbus 340 family, the fly-by-wire system, which is an extremely safety-critical feature, incorporates multiple redundancy [112]. There are three primary and two secondary main computers, each one comprising two units with different software. The primary and secondary computers run on different processors, and have different hardware and different architectures. They were designed and are supplied by different vendors. Only one flight computer is sufficient for full operation. Since mechanical signaling was retained for rudder movement and horizontal stabiliser trim, the aircraft can, if necessary, still be flown relying on mechanical systems only. Each computer has its command and monitoring units running in parallel; see Figure 1.2. They have separate hardware. The software for different channels in each computer was designed by different groups using different languages. Each control surface is controlled by different actuators which are driven by different computers. The hydraulic system is triplicated and the corresponding lines take different routes through the aircraft. The power supply sources and the signaling lanes are segregated.

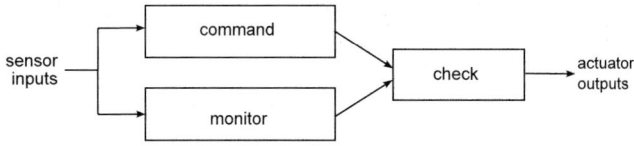

Fig. 1.2. Architecture of an A340 computer

In case of a loss of system resources, the flight control system may be reconfigured dynamically. This involves switching to alternative control software while maintaining system availability. Three operational modes are supported:

Normal - control plus reduction of workload,
Alternate - minimum computer-mediated control, and
Direct - no computer-mediation of pilot commands.

In spite of all these measures, there has been a number of incidents and accidents that may be related to the flight control system or its specifications, although a direct dependence has never been proven.

As functional and non-functional demands for computer systems have continued to grow over the last 30 years, so has the size of the resulting systems. They have become extremely large, consisting of many components, including distributed and parallel software, hardware, and communications, which increasingly interface with a large number of external devices, such as sensors and actuators. Another reason for large (and certain small) systems growing extensively complex is also the large number and complexity of interconnections between their components. Naturally, neither size nor number of connections nor components are the only sources of complexity. As users place increasing importance on such non-functional objectives as availability, fault tolerance, security, safety, and traceability, the operation of a complex computer system is also required to be "non-stop", real-time, adaptable, and dependable, providing graceful degradation.

It is typical that such systems have lifetimes measured in decades. Over such periods, components evolve, logical and physical interconnections change, and interfaces and operational semantics do likewise, often leading to increased system complexity. Other factors that may also affect complexity are geographic distribution of processing and databases, interaction with humans, and unpredictability of system reactions to unexpected sequences of external events. When left unchecked, non-functional objectives, especially in legacy systems, can easily be violated. For instance, there are big, commercial off-the-shelf, embedded systems now running large amounts of software basically unknown to the user, which are problematic when trying to use them for real-time applications.

The safety of control systems needs to be established by certification. In that process, developers need to convince official bodies that all relevant hazards have been identified and dealt with. Certification methods and procedures used in different countries and in different industry domains vary to a large extent. Depending on national legislation and practice, currently the licensing authorities are still very reluctant or even refuse to approve safety-related technical systems, whose behaviour is exclusively program-controlled. In general, safety-licensing is denied for highly safety-critical systems relying on software with non-trivial complexity. The reasons lie mainly in a lack of confidence in complex software systems, and in the considerable effort needed for their safety validation. In practice, a number of established methods and guidelines have already proven its usefulness for the development of high integrity software employed for the control of safety-critical technical processes. Prior to its application, such software is further subjected to appropriate measures for its verification and validation.

However, according to the present state-of-the-art, all these measures cannot guarantee the correctness of larger programs with mathematical rigour. The method of diverse back-translation, for instance, which is the only general method approved by TÜV Rheinland (a public German licensing authority) to verify safety-critical software, is so cumbersome that up to two person-months are needed to verify just 4kB of machine code [48]. Practice has shown that even such small software components may include severe deficiencies as software developers mainly focus on functionality and often neglect safety issues. The problems encountered are exacerbated by the need to verify proper real-time behaviour.

1.3.1 Brief History of Safety Standards Relating to Computers in Control

This section provides a brief historical overview of the most important international, European and German safety standards. The list is roughly ordered by the year of publication.

DIN V VDE 0801 and DIN V 19250: [31, 32]

These documents belong to the first German safety standards applicable to general electric/electronic/programmable electronic (E/E/PE) safety-related systems comprehensively covering software aspects. Previous standards that dealt with the use of software covered only few life-cycle activities and were rather sector-specific, *e.g.*, IEC 60880 [57] "Software for Computers in the Safety Systems of Nuclear Power Stations". Although officially published in different years, *viz.*, DIN V VDE 0801 in 1990 and DIN V 19250 in 1994, there is a close link between them. They establish eight safety requirement classes (German: **A**nforderungs**k**lassen), with AK 1 the lowest and AK 8 the highest.

DIN V VDE 0801: Principles for using Computers Safety-related Systems

This standard defines techniques and measures required to meet each of the requirement classes. It includes techniques to control the EFFECT of hardware failures as well as measures to avoid the insertion of design-faults during hardware and software development. These measures cover design, coding, implementation, integration and validation, but the life-cycle approach is not explicitly mentioned.

DIN V 19250: Control Technology; Fundamental Safety Aspects for Measurement and Control Equipment

This standard specifies a methodology to establish the potential risk to individuals. The methodology takes the consequences of failures as well as the their probabilities into account. A risk graph is used to map the potential risk to one of the eight requirement classes.

EUROCAE-ED-12B: Software Considerations in Airborne Systems and Equipment Certification [38]

This standard, which is equivalent to the US standard RTCA DO-178B, was drafted by a co-operation of the European Organisation for Civil Aviation Equipment (EUROCAE) and its US counterpart Radio Technical Commission for Aeronautics (RTCA). It was released in 1992 and replaces earlier versions published in 1982 (DO-178/ED-12) and in 1985 (DO-178A/ED-12A). The standard considers the entire software life-cycle and provides a thorough basis for certifying software used in avionic systems like airplanes. It defines five levels of criticality, from A (Software whose failure would cause or contribute to a catastrophic failure of the aircraft) to E (Software whose failure would have no effect on the aircraft or on pilot workload).

EN 954: Safety of Machinery — Safety-related Parts of Control Systems [37]

This standard was developed by the European Committee for Standardisation (CEN) and has two parts: *General Principles for Design* and *Validation, Testing, Fault Lists.* Part 1 was first released in 1996, Part 2 in 1999. The standard complies with the basic terminology and methodology introduced in EN 292-1 (1991), and covers the following five steps of the safety life-cycle: hazard analysis and risk assessment, selection of measures to reduce risk, specification of safety requirements that safety-related parts must meet, design, and validation. It defines five safety categories: B, 1, 2, 3 and 4. The lowest category is B which requires no special measures for safety, and the highest is 4 requiring sophisticated techniques to avoid the consequences of any single fault. The standard focuses merely on the application of fault tolerance techniques in parts of machinery, it does not consider the system and its life-cycle as a whole [102].

ANSI/ISA S84.01: Application of Safety Instrumented Systems for the Process Industry [3]

This is the US standard for safety systems in the process industry. It was primarily introduced in 1996, and founded on the draft of IEC 61508 published in 1995. The standard follows nearly the same life-cycle approach as IEC 61508 and, thus, can be considered a sector-specific derivative of this umbrella standard. The specialisation on the process industry becomes apparent by its strong focus on *Safety Instrumented Systems* (SIS) and *Safety Instrumented Functions* (SIFs). According to the standard, SISs transfer a process to a safe state in case predefined conditions are violated, such as overruns of pressure or temperature limits. SIFs are the actions that a SIS carries out to achieve this. Since the committee initially thought that SIL 4 applications do not exist in the process industry, the first edition defined only three SILs, which are equivalent to SIL 1 to 3 of IEC 61508. However, the new release, ANSI/ISA S84.00.01-2004, includes the highest class SIL 4.

IEC 61508: Functional Safety of Electrical/Electronic/Programmable Electronic (E/E/PE) Safety-related Systems [58]

The first draft of this standard was devised by IEC's Scientific Committee 65A and published in 1995 under the name "IEC 1508 Functional Safety: Safety-related Systems". After it gained wide publicity, a revised version was released in December 1998 as IEC 61508. This version comprises seven parts:

Part 1: General requirements

Part 2: Requirements for electrical/electronic/programmable electronic safety-related systems

Part 3: Software requirements

Part 4: Definitions and abbreviations

Part 5: Examples of methods for the determination of safety integrity levels

Part 6: Guidance on the application of IEC 61508-2 and IEC 61508-3

Part 7: Overview of techniques and measures

The first four parts are normative, *i.e.*, they state definite requirements, whereas Parts 5 to 7 are informative, *i.e.*, they supplement the normative parts by offering guidance rather than stating requirements.

The standard defines four Safety Integrity Levels (SILs). SIL 1 is the lowest, SIL 4 the highest safety class. It is important to note that SILs are measures of the safety requirements of a given process; an individual product cannot carry a SIL rating. If a vendor claims a product to be certified for SIL 3, this means that it is certified for use in a SIL 3 environment [102].

The standard has a "generic" character, *i.e.*, it is intended as basis for writing sector- or application-specific standards. Nevertheless, if application-specific standards are not available, this umbrella standard can be used on its own.

In December 2001, CENELEC published a European version as EN 61508. It obliged all its member countries to implement this European version at national level by August 2002, and to withdraw conflicting national standards by August 2004. That is why DIN V VDE 0801 and DIN V 19250, as well as their extensions, were withdrawn at that date.

EN 50126, EN 50128 and EN 50129: CENELEC railway standards
[34, 35, 36]

These three standards represent the backbone of the European safety licensing procedure for railway systems. They were developed by the Comité Européen de Normalisation Electrotechnique (CENELEC), the European Committee for Electrotechnical Standardisation in Brussels.

EN 50126: Railway Applications — The Specification and Demonstration of Dependability, Reliability, Availability, Maintainability and Safety (RAMS)

EN 50128: Railway Applications — Software for Railway Control and Protection Systems

EN 50129: Railway Applications — Safety-Related Electronic Systems for Signaling

This suite of standards, which is often referred to as the "CENELEC railway standards", was created with the intention to increase compatibility between rail systems throughout Europe and to allow mutual acceptance of approvals given by the different railway authorities. EN 50126 was published in 1999, whereas EN 50128 and EN 50129, which represent application-specific derivatives of IEC 61508 for railways, were released in 2002.

IEC 61511: Functional Safety: Safety Instrumented Systems for the Process Industry Sector [59]

This safety standard was first released in 2003, and represents a sector-specific implementation of IEC 61508 for the process industry. Thus, it covers the same safety life-cycle approach and re-iterates many definitions of its umbrella standard. Aspects that are of crucial importance for this application area, such as sensors und actuators, are treated in considerably higher detail. The standard consists of three parts named "Requirements", "Guidance to Support the Requirements", and "Hazard and Risk Assessment Techniques".

In September 2004, the IEC added a "Corrigendum" to the standard, and the ANSI adopted this version as new ANSI/ISA 84.00.01-2004 (IEC 61511 MOD). The US version is identical to IEC 61511 with one exception, a "grandfather clause" that preserves the validity of approvals for existing SISs.

IEC 61513: Nuclear Power Plants — Instrumentation and Control for Systems Important to Safety — General Requirements for Systems [60]

This sector-specific derivative of IEC 61508 for nuclear power plants was primarily released in 2002. Other safety standards for nuclear facilities like, e.g., IEC 60880 were revised in conformity with IEC 61508.

There are many more safety standards related to Programmable Electronic Systems (PES), especially in the military area. This sometimes causes uncertainty in choosing the standard applicable for a given application, e.g., EN 954-1 or IEC 61508 [41]. Moreover, if a system is used in several regions with different legal licensing authorities, e.g., intercontinental aircraft, they may need to conform with multiple safety standards.

The overview presented in this section highlights the importance of IEC 61508. Its principles are internationally recognised as fundamental to modern safety management. Its life-cycle approach and holistic system view is applied in many modern safety standards — not only the ones that fall under the regulations of CENELEC.

1.3.2 Safety Integrity Levels

In the late 1980s, the IEC started the standardisation of safety issues in computer control [58]. They identified four Safety Integrity[2] Levels SIL 1 to SIL 4, with SIL 4 being the most critical one. In Table 1.1, applicable programming methods, language constructs, and verification methods are assigned to the safety integrity levels.

---

[2] Safety integrity is the likelihood of a safety-related system to perform the required safety functions satisfactorily under all stated conditions within a stated period of time [107].

Table 1.1. Safety integrity levels

Safety integrity level	Verification method	Language constructs	Typical programming method
SIL 4	Social consensus	Marking table entries	Cause-effect tables
SIL 3	Diverse back translation	Procedure calls	Function block diagrams with formally verified libraries
SIL 2	Symbolic execution, formal correctness proofs	Procedure call, assignment, case selection, iteration restricted loop	Language subsets enabling (formal) verification
SIL 1	All	Inherently safe ones, application oriented ones	Static language with safe constructs

For applications with highest safety-criticality falling into the SIL 4 group, one is not allowed to employ programming means such as we are used to. They can only be "programmed" using cause-effect tables (such as programming of simple PLA[3], PAL and similar programmable hardware devices), which are executed by hardware proven correct. The rows in cause-effect tables are associated with events, occurrence of which gives rise to Boolean preconditions. They can be verified by deriving the control functions from the rules read out from the tables stored in permanent memory and comparing them with the specifications. In Figure 1.3 a safety-critical fire fighting application is presented as a combination of cause-effect tables and functional block macros.

At SIL 3, programming of sequential software is already allowed, although only in a very limited form as interconnection of formally verified routines. No compilers may be used, because there are no formally proven correct compilers yet. A convenient way to interconnect routines utilises Function Block Diagrams as known from programmable logic controllers [56]. The suitable verification method is diverse back-translation: several inspectors take a program code from memory, disassemble it, and derive the control function. If they can all prove that it matches the specifications, a certificate can be issued [73]. This procedure is very demanding and can only be used in the case of pre-fabricated and formally proven correct software components.

---

[3] Programmable Logic Array.

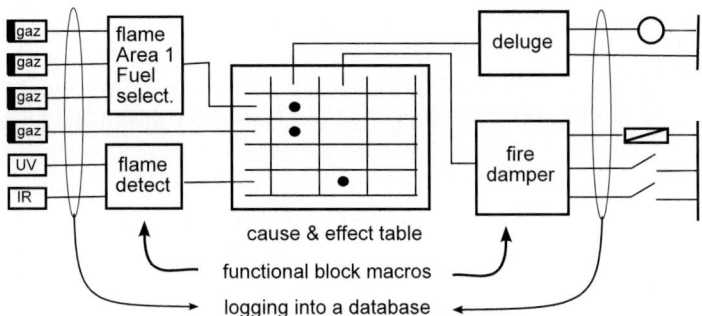

Fig. 1.3. An example of a safety-critical application

SIL 2 is the first level to allow for programming in the usual sense. Since formal verification of the programs is still required, only a safe subset of the chosen language may be used, providing for procedure calls, assignments, alternative selection, and loops with bounded numbers of iterations.

Conventional programming is possible for applications with the integrity requirements falling into SIL 1. However, since their safety is still critical, only static languages are permitted without dynamic features such as pointers or recursion that could jeopardise their integrity. Further, constructs that could lead to temporal or functional inconsistencies are also restricted. Any reasonable verification methods can be used.

In this book, applications falling into SIL 1 will be considered, although for safety back-up systems or partial implementations of critical subsystems higher levels could also apply. For that reason, in the sequel we shall only refer to SIL 1.

1.3.3 Dealing with Faults in Embedded Control Systems

A good systematic elaboration of handling faults and a taxonomy from this domain was presented by Storey [107]. Some points are summarised below. Faults may be characterised in different ways, for example, by:

Nature: random faults (hardware failure), systematic faults (design faults, software faults);

Duration: permanent (systematic faults), transient (alpha particle strikes on semiconductor memories), intermittent (faulty contacts); or by

Extent: local (single hardware or software module), global (system).

More and more, the general public is realising the inherent safety problems associated with computerised systems, and particularly with their software. Hardware is subject to wear, transient or random faults, and unintended environmental influences. These sources of non-dependability can, to a very large extent, be coped with successfully by applying a wide spectrum of redundancy and fault-tolerance methods.

Software, on the other hand, does not wear out nor can environmental circumstances cause software faults. Instead, software is imperfect, with all errors being design errors, *i.e.*, of systematic nature, and their causes always being latently present. They originate from insufficient insight into the problems at hand, leading to incomplete or inappropriate requirements and design flaws. Programming errors may add new failure modes that were not apparent at the requirements level. In general, not all errors contained in the resulting software can be detected by applying the methods prevailing in contemporary software development practice. Since the remaining errors may endanger the environment and even human lives, embedded systems are often less trustworthy than they ought to be. Taking the high and fast increasing complexity of control software into account, it is obvious that the problem of software dependability will exacerbate severely.

As already mentioned, due to the complexity of programmable control systems, faults are an unavoidable fact. A discipline coping with them is called "fault management". Broadly, its measures can be subdivided into four groups of techniques:

Fault avoidance aims to prevent faults from entering a system during the design stage,

Fault removal attempts to find faults before the system enters service (testing),

Fault detection aims to find faults in the system during service to minimise their effects, and

Fault tolerance allows the system to operate correctly in the presence of faults.

The best way to cope with faults is to prevent them from occurring. A good practice is to restrict the use of potentially dangerous features. Compliance with these restrictions must be checked by the compiler. For instance, dynamic features like recursion, references, virtual addressing, or dynamic file names and other parameters can be restricted, if they are not absolutely necessary.

It is important to consider the possible hazards, *i.e.*, the capability to do harm to people, property or the environment [107], during design time of a control system. In this sense the appropriate actions can be categorised as:

- Identification of possible hazards associated with the system and their classification,
- Determination of methods to dealing with these hazards,
- Assignment of appropriate reliability and availability requirements,
- Determination of an appropriate Safety Integrity Level, and
- Specification of appropriate development methods.

Hazard analysis presents a range of techniques that provide diverse insight into the characteristics of a system under investigation. The most common approaches are Failure Modes and Effects Analysis (FMEA), Hazard and Operability Studies (HAZOP), and the Event- and Fault Tree Analyses (ETA and FTA).

Fault tree analysis in particular appears to be most suitable for use in the design of embedded control systems. It is a graphical method using symbols similar to those used in digital systems design, and some additional ones representing primary and secondary (the implicit) fault events to represent the logical function of the effects of faults in a system. The potential hazards are identified; then the faults and their interrelations that could lead to undesired events are explored. Once the fault tree is constructed it can be analysed, and eventually improvements proposed by adding redundant resources or alternative algorithms.

Since it is not possible in non-trivial cases to guarantee that there are no faults, it is important to detect them properly in order to deal with them. Some examples of fault-detection schemes are:

Functionality checking involves software routines that check the functionality of the hardware, usually memories, processor or communication resources.

Consistency checking. Using knowledge about the reasonable behaviour of signals or data, their validity may be checked. An example is range checking.

Checking pairs. In the case of redundant resources it is possible to check whether different instances of partial systems behave similarly.

Information redundancy. If feasible, it is reasonable to introduce certain redundancy in the data or signals in order to allow for fault detection, like checksums or parities.

Loop-back testing. In order to prevent faults of signal or data transmission, they can be transmitted back to the sources and verified.

Watchdog timers. To check the viability of a system, its response to a periodical signal is tested. If there is no response within a predefined interval, a timer detects a fault.

Bus monitoring. Operation of a computer system can often be monitored by observing the behaviour on its system bus to detect hardware failures.

It is advisable that these fault-detection techniques are implemented as operating system kernel functions, or in any other way built into the system software. Their employment is thus technically decoupled from their implementation allowing for their systematic use.

1.3.4 Fault-tolerance Measures

Approaches of Fault-Tolerant Control can be divided into two categories: passive and active fault tolerant control. The key difference between them is that an active fault tolerant control system includes a fault detection and isolation (FDI) system, and that fault handling is carried out based on information on faults delivered by the FDI system. In a passive fault tolerant control system, on the other hand, the system components and controllers are designed to be robust to possible faults to a certain degree. Figure 1.4 sketches the basic classification of fault tolerant control concepts.

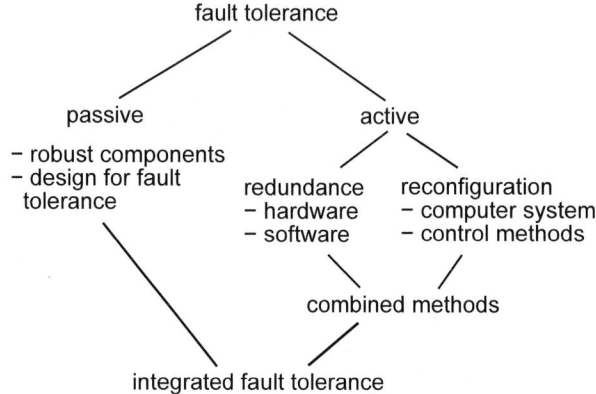

Fig. 1.4. Classification of fault-tolerance measures

Passive measures to improve fault tolerance mean that any reasonable effort must be made to make a design robust. For instance, the components must be selected accordingly, and with reasonable margins in critical features. Also, fault tolerance should already be considered in the design of subsystems. In addition to enhancing the quality and robustness of process components, using redundancy is a traditional way to improve process reliability and availability. However, because of the increased costs and complexity of the system, its usability is limited.

Evidently more flexible and cost effective is the reconfiguration scheme. Fault tolerance is achieved by system and/or controller reconfiguration, *i.e.*, after faults are identified and a reduction of system performance is observed, the overall system performance will be recovered (possibly to an acceptable degree, only) by a reconfiguration of parts of the control system under real-time conditions. This is a new challenge in the field of control engineering. In the following, the most common approaches for this are briefly sketched.

Redundancy

The most common measure to make a system tolerant to faults is to employ redundant resources. In the area of computing this idea originated in 1949: although still not tolerant to faults, EDVAC already had two ALUs to detect errors in calculation. Probably the first fault-tolerant computer was SAPO [87] built in Prague from 1950 to 1954 under the supervision of A. Svoboda, using relays and a magnetic drum memory. The processor used triplication and voting, and the memory implemented error detection with automatic retries when an error was detected. A second machine developed by the same group (EPOS) also contained comprehensive fault-tolerance features.

The most simple model of redundancy-based fault tolerance is Triple Modular Redundancy (TMR): three resources with a voter allow for recognising a single resource failure. This is also called two-out-of-three ($2oo3$). The model can be extended to N-Modular Redundancy (NMR), which can accommodate more failures, but becomes more and more expensive. Seldom does N exceed 5 ($3oo5$). For other examples please refer to Section 7.3.1.

There are several ways in which the redundancy can be employed:

Hardware redundancy: e.g., in form of TMR,
Software redundancy: e.g., diversely implemented routines in recovery blocks, see below,
Information redundancy: e.g., as parity bits or redundant codes, and
Time redundancy: i.e., repetitive calculations to cope with the intermittent faults. fault-detection

Both in hardware and software redundancy it is most important to avoid common mode failures. They can be coped with successfully by design diversity. For a good example, see the description of the Airbus fly-by-wire computer control system on Page 14. To achieve a fully decoupled design, one should start with separate specifications to avoid errors in this stage, which are most costly and difficult to find by testing — a wrong system can thus even be verified formally!

Sometimes, the sources of common mode faults are deeply anchored in the designers by their education and training. This introduces a social component of fault management. This social problem of common mode failures was addressed by M.P. Hilsenkopf, an expert concerned with research and development for French nuclear power plants [53]. He pointed out that safety critical components are developed in parallel in two places by two different groups without a contact, *viz.*, in Toulouse and Paris. Starting with separate acquisition of requirements from physicists and engineers, they each provide their own specifications each. Based on that, they develop, verify and validate two different solutions. Then they swap them for final testing. Since they have both specified and developed their solutions, both groups know the problem very well, they know which questions and difficulties they had to solve, and they check how the respective competing group has coped with them. Thus, they verify each others' design on their own specifications, and both try to prove that their solution is better.

Reconfiguration

Owing to the fixed structure and high demands for hardware and software resources, and for economic reasons in less critical applications, the employment of the redundancy strategy is usually limited to certain specific technical processes or to key system components like the central computing system or the bus system. To cover entire systems, fault-tolerant strategies with fault accommodation or system and/or controller reconfiguration are more suitable.

Owing to significantly different functions and working principles, the problems related to reconfiguration of control methods, algorithms and approaches as well as of the computer platforms, on which the latter and signal and data processing are running, are generally considered in separate and usually independent contexts. In this book we shall deal in more detail with the aspect of reconfiguring computer control systems, and with supporting the methods of higher-level control system reconfiguration.

When the occurrence of faults is detected, a system decides either to accommodate these faults or to perform controller and/or hardware reconfiguration, which might imply graceful performance degradation or, in the case of severe faults, to drive the component or process concerned into a safe mode. The decision on the type of action taken by the reconfiguration system is based on the evaluation of the actual system performance provided by a monitoring system, and on the need to assure an adequate transient behaviour upon reconfiguration.

Thus, for reconfiguration, designers prepare different solutions to a problem with gradually decreasing quality; the last one usually drives the system into a safe failing mode. For each of the solutions the necessary resources and the preconditions are listed. On the other hand, the computer system architecture provides for reconfiguration of the hardware in order to cope with resource failure. The optimum approach is normally run at the start, utilising all (or most of) the computer and control resources. This is also the main difference between redundancy and reconfiguration: most of the operational resources are always used.

Important components of reconfiguration-based fault-tolerant control are a fault detection system and a reconfiguration manager: based on the information provided by the former and on the resource needs attached to the gradually degraded options, the latter decides on the operating mode. Since the reconfiguration manager represents a central point of failure, it must be implemented in the safest possible way. It may be redundant or distributed among the other resources, which by a voting protocol then compete for taking over control. An example for the latter solution is the protocol to select a new time master in the case of failure in the TTCAN protocol [64]; cf. Section 3.5.2.

With respect to application design, dynamic reconfiguration resembles software redundancy (see below) — the recovery approaches or N versions of a control algorithm. According to the occurrence of faults, control will be switched to the most competent version, utilising the sound resources, and providing the best possible quality of service.

Hardware Fault Tolerance

In the design of fault-tolerant hardware, systems can be classified as static, dynamic or hybrid.

Static systems utilise fault effect masking, *i.e.*, preventing the effects of faults to propagate further into the systems. They use some voting mechanism to compare the outputs and mask the effect of faults. Well-known approaches include the n-modular redundancy, TMR with replicated voters (to eliminate the danger that a single voter fails).

Dynamic systems try to detect faults instead of masking their effects. The technique used is providing the stand-by resources. There is a single system producing output. If a fault is detected, control is switched to a redundant resource. The latter may have been running all the time in parallel with the master resource (hot stand-by), or was switched on when the master resource failed. In the case of hot redundancy, faults can be detected by, *e.g.*, self-checking pairs. In this case, obviously the redundant resource is more reliable, although probably delivering lower quality of service.

Static redundancy masks faults and they do not appear in the system. This is, however, achieved at a high cost for massive redundancy. Dynamic systems utilise less redundancy and provide continuous service, but introduce transient faults to the system: when an error occurs, the fault is in the first instant propagated, then coped with.

Hybrid systems combine the properties of the former two classes: by combining the techniques of dynamic redundancy with voters, they detect faults, mask their effects, and switch to stand-by units.

Software Fault Tolerance

Software fault tolerance copes with faults (of whatever origin) by software means rather than tolerating software errors. There are two major methods falling into this category, *N-version programming* and *recovery blocks*.

N-version programming resembles static hardware redundancy. A problem is solved by N different and diversely designed programs, all fulfilling the requirements stated in the specifications. All these programs are then run on the same data, either sequentially, or in parallel in the case of a multiprocessing system. If the results do not match, a fault is assumed and the results are blocked. The disadvantage of this method is that it either requires a multiprocessor system or takes more than N-times more time even if the system is working correctly.

Recovery blocks do not take much more time if there is no fault. Again, N versions of a program are designed, but only the most appropriate one is executed first. If a fault can be recognised, the results are discarded, the intermediate system states are reset, and the next alternative solution is tried. Eventually, one alternative must fulfil the requirements, or at least assure failsafe behaviour. There are two different types of recovery blocks techniques [23, 24], backward and forward recovery.

The principle of *backward recovery* is to return to a previous consistent system state after an inconsistency (fault) is detected by consistency tests called postconditions. This can be done in two ways; (1) by the operating system recording the current context before a program is "run", and restoring it after its unsuccessful termination, or (2) by recovery blocks inside the context of a task. The syntax of a recovery block *(RB)* could be

$$\text{RB} \equiv \textbf{ensure } post \textbf{ by } P_0 \textbf{ else by } P_1 \textbf{ else by } \ldots \textbf{else } failure$$

where P_0, P_1, *etc.* are alternatives which are consecutively tried until either consistency is ensured by meeting the *post*-condition, or the *failure*-handler is executed to bring the system into a safe state. Each alternative should independently be able to ensure consistent results.

The *forward error recovery technique* tries to obtain a consistent state from partly inconsistent data. Which data are usable can be determined by consistency tests, error presumptions, or with the help of independent external sources.

If in embedded process control systems an action directly affects a peripheral process, an irreversible change of initial states inside a failed alternative is caused. In this case, backward recovery is generally not possible. As a consequence, no physical control outputs should be generated inside the alternatives which may cause backward recovery in case of failure, *i.e.*, inside those which have postconditions. If this is not feasible, only forward recovery is possible, bringing the system into a certain predefined, safe, and stable state.

1.4 Summary of Chapter 1 and Synopsis of What Follows

In this introduction, some definitions have been laid down and the nature of real-time systems' execution behaviour has been demonstrated on examples. Safety issues have been discussed together with a brief enumeration of existing fault management measures and standards. This chapter has introduced terminology and provided the structure of guidelines on which the remaining chapters will build.

In the remainder of Part I, the concepts most important in designing distributed embedded real-time control systems will be elaborated. To start with, multitasking is the topic of Section 2, as it presents the nature of complex embedded control systems. In the section, first the approaches to task management are introduced and, then, two most common issues, scheduling and synchronisation, are dealt with. The preferred solutions are elaborated in more detail.

Sections 3 and 4 more specifically deal with hardware and software aspects of embedded systems design. They present some original solutions to certain problems, which have been developed by the authors, and which constitute guidelines for the implementation of platforms for distributed systems. In Part II the latter are elaborated in detail.

2

Multitasking

Tasks (or computing processes) are the smallest entities and fundamental elements of asynchronous programming. They represent the execution of programs in a computing system during the lifetime of the processes. Such a lifetime begins with the registration instant of a process and ends with its cancellation. Thus, the computing process exists not only during the actual processing in a functional unit of a computing system, but also before the start of processing and during the planned and forced waiting periods.

Tasks are programmed using adequate programming methods, languages and tools. The basic requirement for hard real-time tasks is that it must be possible to assess their maximum execution time, called Worst-Case Execution Time (WCET). A precondition, in turn, for this is that the behaviour of the execution platform is deterministic and predictable. More details about WCET analysis will be given in Section 4.2.

In the following sections, first the concepts of task and multitasking will be dealt with. Then, the ultimate guideline to design architectures for embedded systems — schedulability — will be explored, and a feasible scheduling policy presented. Finally, the principles of inter-task synchronisation will be explained.

2.1 Task Management Systems

Execution of tasks is administered by system programs. Since there are (except in very specific applications) never enough processing units that each task could exclusively be mapped on one of its own, tasks must be arranged or *scheduled* in a sequence for execution. Sometimes, tasks can be scheduled *a priori* during design time. This is only possible if the system behaviour is completely known in advance like, *e.g.*, for a washing machine, whose tasks filling with water, heating, turning the drum either way, draining water, *etc.* follow sequences that are always the same. Such systems are called static. Here, the execution schedules of parallel tasks can be carefully planned, optimised, and verified off-line during design time. Consequently, this is the safest way to implement embedded applications.

Most systems, however, are dynamic: at least some of the events requiring service by tasks are known in advance with their parameters, but their occurrence instants are sporadic and asynchronous to other control system processings. In such cases, tasks need to be scheduled at run-time. These systems are more flexible, but their design is also much more challenging. In this chapter we shall deal with dynamic or mixed systems. There are different approaches to how mixtures of static and dynamic tasks can be executed. Two representative examples are the *cyclic executive approach* and *asynchronous multitasking*.

2.1.1 Cyclic Executive

The simpler approach of the two can accommodate both statically scheduled periodic tasks and asynchronous sporadic tasks, although for the latter it can only be used with limitations. The approach is based on periodic execution of a schedule, part of which is determined statically and the other part dynamically. The start of each cycle is triggered by a timer interrupt.

Periodic tasks may have different requirements and need to be executed within their specified periods. Further, it is assumed that the tasks are completely independent of each other. The periodic tasks are assigned to be run in cycles according to their requirements. The longest cycle is called the *major cycle*, consisting of a number of *minor cycles* of different durations. All process periods must be integer multiples of the minor cycle.

It is possible to include asynchronously arriving sporadic tasks into this scheme: they can be executed in the slack time between the termination of the last periodic task in a cycle and the start of the next cycle. This also means that the execution time of a sporadic task is limited to this slack; if it is longer, an overrun situation occurs. A major use of such sporadic tasks is to report exceptions.

An example of a schedule generated by a cyclic executive for periodic tasks A, B, C and D with periods 1, 2, 2 and 4, respectively, and two sporadic tasks E1 and E2 is shown in Figure 2.1.

A very simple implementation of the cyclic executive approach is sketched by pseudo-code in Figure 2.2. It is based on an array of task control blocks. For each task there are parameters: duty cycle, cycle counter, and start address. Another array mentions signals and addresses of sporadic tasks. It is assumed that the worst-case execution times are shorter than the smallest slack between the end of the last periodic task and the beginning of the next cycle, which can be calculated at design time. Further, it is assumed that sporadic tasks are rare; if there are more in a queue they are executed in first-in-first-out (FIFO) manner. Finally, there is a check at the beginning whether it came to an overload in the preceding cycle; that may happen if the worst-case execution times of tasks were exceeded.

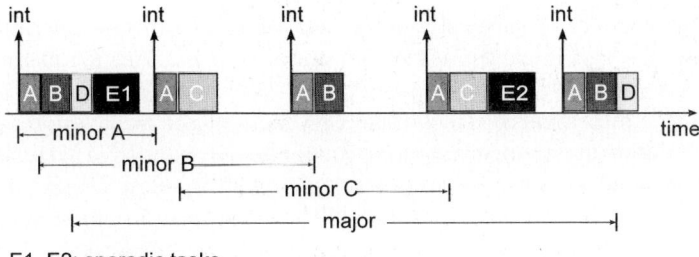

E1, E2: sporadic tasks

Fig. 2.1. A schedule generated by a cyclic executive

```
on timer interrupt:
        IF NOT waiting_for_interrupt THEN
            break execution;
            overrun error
        END IF;
        waiting_for_interrupt:=false;
        FOR i:= 1 TO number_of_periodic_tasks DO
            cycle_counter[i]:=cycle_counter[i]+1;
            IF cycle_counter[i]=duty_cycle[i] THEN
                cycle_counter[i]:=0;
                execute periodic_task[i]
            END IF
        END DO;
        FOR i:= 1 TO number_of_possible_signals DO
            IF signal[i] has arrived THEN
                execute asynchronous_task[i]
            END IF
        END DO;
        waiting_for_interrupt:=true;
        wait for interrupt
```

Fig. 2.2. Algorithm of cyclic executive

It must be noted that the "tasks" here are actually only procedure calls. No usual tasking operations can be performed; their static schedule assures them exclusive execution slots. Thus, there is no interdependency, and synchronisation with its possible dangers is not necessary. Execution of the tasks is temporally predictable in full. If it is difficult to construct a feasible schedule off-line due to high system load, different optimisation methods can be utilised. Once a schedule is composed, no further run-time schedulability tests are needed for periodic tasks. In the most simple, although not very rare cases, the cyclic executive reduces to periodic invocation of a single task. Simple programmable logic controllers (PLC) operate in this manner, first acquiring sensor inputs, then executing control functions, and finally causing process actuations.

A large number of successful implementations of process control application using the cyclic executive can be found. Lawson [76], for instance, has
presented a philosophy, paradigm and model called *Cy-Clone* (Clone of Cyclic
Paradigm), which aims to remove the drawbacks of the cyclic approach while
providing deterministic properties for time- and safety-critical systems. In his
cyclic approach, the period should be long enough to allow for all processing,
but sufficiently short to ensure the stability of control. To adapt to dynamic
situations, the method can provide mode shifts, where the period varies in
a controlled manner during execution. As Lawson points out, the Cy-Clone
approach is not claimed to be the best solution for all real-time systems, but
was established on the basis *"If the shoe fits, wear it"*. In the early 1980s it was
employed in a control system of the Swedish National Railway to assist train
engineers in following the speed limits along the railway system in Sweden. In
2000, Lawson reported [77] that it was still working (after a non-significant
re-design in 1992), but now in high-speed trains!

2.1.2 Asynchronous Multitasking

Although convenient and safe, the above approach cannot be used for dynamic systems, where all or most tasks are asynchronous. In such cases, real
asynchronous multitasking needs to be employed. In contrast to the static
cyclic executive, timeliness and simultaneity are to be provided dynamically.
The dynamic load behaviour requires scheduling problems to be solved with
an adequate strategy. The demand for simultaneity implies *pre-emptability*
of tasks: a running task may be pre-empted and put back into the waiting
state if another task needs to be executed according to the scheduling policy.
However, as a consequence this leads to synchronisation requirements.

Context-switching, *i.e.*, swapping the tasks that are executing on a processor, starts with saving the context (*i.e.*, the state consisting of program
counter, local variables, and other internal task information) of the executing task to storage in order to allow its eventual resumption when it is re-
scheduled to run. This storage is often called Task Control Block, (TCB) and
also includes other information about tasks, *e.g.*, their full and remaining execution time, deadlines, resource contention, *etc.* Then, the state of the newly
scheduled task is loaded and the task is run. Obviously, as this represents a
non-productive overhead consuming certain processing capacity, it should be
kept to a minimum by choosing the most adequate task scheduling policy.

Multitasking must be supported by operating systems. Any one mainly
supports its own scheme of tasks states and transitions or has, at least, different names for them. A typical, simple but sufficient task state transition
diagram is presented in Figure 2.3. In this model, tasks may assume one of
the following states:

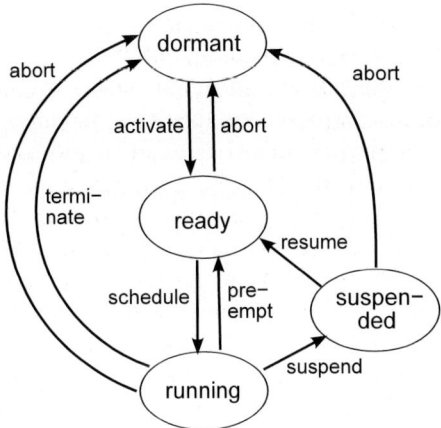

Fig. 2.3. Task state transition diagram

Dormant (also known or terminated). Once a task is introduced in a system by its initialisation, it enters the dormant state. Here it is waiting to be activated by an event. Further, when the task accomplishes its mission and *terminates* in the "running" state, it is brought back to the dormant (or, in this case better, terminated) state. If for any reason a task is not needed any more in any other state, it can be *aborted*, *i.e.*, put into the dormant state again.

Ready. A task in the dormant state can be *activated* by an event and then joins the set of tasks ready to be scheduled for execution. Such a task enters the ready state. A task can also enter this state when it is *pre-empted* from running, or *resumed* from suspension. Upon occurrence of any event changing the set of ready tasks, a scheduling algorithm is invoked to order them, according to a certain policy, in a sequence in which they will be executed. If a task from the ready set has lost its meaning for any reason (*e.g.*, it was explicitly aborted by another task, or its waiting time for resources has exceeded, so that its deadlines cannot be met anymore) it can be *aborted* to the dormant state.

Running. As a result of *scheduling* the tasks in the ready state, the first task in the sequence is dispatched to the running state, *i.e.*, to be run on the processor. Eventually, it is *terminated*, when finished, or *aborted* to the dormant state. A running task can be *pre-empted*, *i.e.*, its execution is discontinued and the processor allocated to another task, and it is brought back into the ready state to re-join the set of ready tasks for further scheduling. Pre-emption is related to certain overheads, so one always strives to minimise the number of pre-emptions to the lowest amount possible.

Suspended. If the running task lacks certain resources for further execution, or if any other precedence relation calls for deferral through a synchronisation mechanism, its execution is *suspended*. Once a certain condition that the suspended task is waiting for is fulfilled, it is *resumed* and, thus, brought into the ready state again. If, on the other hand, due to excessive waiting, the task cannot meet its deadline, it is aborted and a certain prepared action is taken in order to deal with this situation.

Task transitions are performed when certain events occur. These may be external events, *e.g.*, signals from the process environment or arrivals of messages, timing events, when a time monitor recognises an instant for triggering a task, or internal events, when, for instance, a semaphore blocks execution of a task. Internal events can also be induced by other tasks. Instead of explicit operating system calls, tasking operations should be supported by adequate language commands. An example of an apt set of tasking commands will be given in Section 4.1.2 in Figure 4.6.

2.2 Scheduling and Schedulability

In Section 1.2.1 it has been established that predictability of temporal behaviour is the most important property of embedded and real-time applications. The final goal is that tasks complete their functions within predefined time periods. Thus, when running a single task it is enough to know its execution time. Dynamic embedded systems, however, must support multitasking. There is always a set of tasks which have been invoked by certain event occurrences, and which compete for the same resources, most notably the processor(s). Since all of them have their deadlines to meet, an adequate *schedule* should be found to assure that. The pre-condition for being able to elaborate all these tasks is, of course, the availability of resources. Lawson [76] presented the concept of *resource adequacy* meaning that there are sufficient resources to guarantee that all processing requirements are met on time.

A schedule to process a task system is called *feasible* if all tasks of the system meet their deadlines (assuming the above mentioned resource adequacy). A task system is called *feasibly executable* if there exists a feasible schedule. At run-time, a *scheduler* is responsible for allocating the tasks ready to execute in a such way that all deadlines are observed, following an apt processor allocation strategy, *i.e.*, a *scheduling algorithm*. A strategy is feasible if it generates for any feasibly executable (free) task set a feasible schedule. Another notion concerning feasibility of a task set is *schedulability*: the ability to find, *a priori*, a schedule such that each task will meet its deadline [108].

Sometimes it may happen that a task cannot meet its deadline; in hard real-time systems this is considered a severe fault. It can be a consequence of improper scheduling, or of system overload. In the former case this is due to selecting an improper or infeasible scheduling strategy. In the latter case the specification of system behaviour and, thus, the dimensioning of the computer control system was wrong. To prevent such specification errors, it is essential to have — prior to system design — a clear understanding of the peak load a system is expected to handle. As it is not possible to design, for instance, a bridge without knowing the load it is expected to carry, so it is not possible to design a real-time system with predictable performance without estimating the load parameters and task execution times.

In order to estimate correctly the necessary capacity of a system and, thus, prevent overload situations, a proof of feasible executability or schedulability of occurring tasks set has to be carried out in the planning phase. This is a difficult problem, especially in dynamic systems with task reacting to sporadically occurring events. There are methods that may provide such proofs, *cf. e.g.*, [108]. Their complexity, however, can easily become NP-complete, and the methods may yield rather pessimistic estimations.

2.2.1 Scheduling Methods and Techniques

Scheduling policies fall into two classes, depending on whether the tasks, once executing, can be pre-empted by other tasks of higher urgency or not. For scheduling dynamic systems, pre-emptions are necessary in order to assure feasibility of the scheduling algorithms. In this subsection, two scheduling policies, fixed priority and rate monotonic scheduling, will be elaborated. Deadline driven scheduling policies, which are problem-oriented and more suitable, will be covered in more detail in the next subsection.

Fixed Priorities

Many popular operating systems, which are currently most frequently employed in embedded systems, base their scheduling on fixed priorities. The advantage is that the latter are in most cases built into the processors themselves in a form of priority interrupt handling systems. Thus, implementation is fast and simple.

As scheduling criteria, however, priorities are not problem-oriented, but artificially invented categories not emerging from the nature of the applications. As a rule, it is difficult to assign adequate priorities to tasks, bearing in mind all their properties (frequency of their invocations, required resources, temporal tightness, relative importance, *etc.*). Designers tend to over-emphasise relative importance, which leads to starvation of other tasks waiting for blocked resources. Priorities are not flexible and cannot adapt to the current behaviour of systems. For these reasons, it is not possible to prove feasibility of such a policy. Once a task is assigned a fixed priority, the scheduler allocates the resources regardless of real needs of that and other tasks.

As an example, let us consider a situation of three tasks with arbitrarily assigned priorities; see Figure 2.4. As we can see, the selection of priorities was unfortunate: the lowest priority task misses its deadlines, although the laxities of the tasks are relatively large. Possibly, the importance of a task suggested it be assigned the highest priority, although its deadline is far and execution

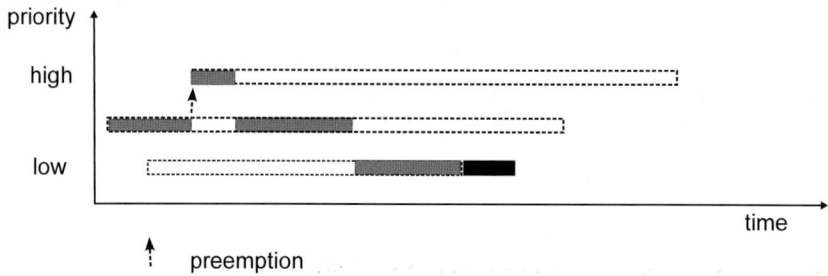

Fig. 2.4. Missing a deadline by inappropriately assigned priorities

time short. A feasible schedule, however, is possible by simply overturning the priorities, as shown in Figure 2.5.

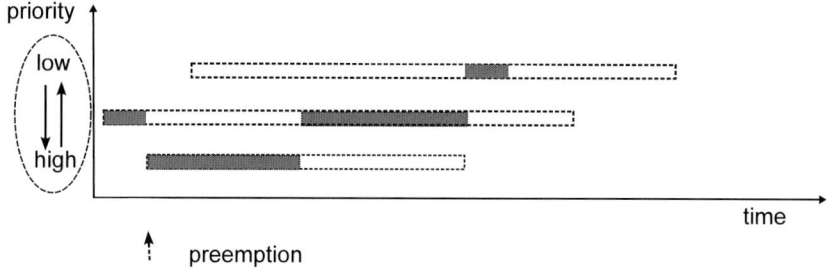

Fig. 2.5. Appropriate assignment of priorities

Further, a number of problems emerge from the priority-based scheduling policy. For instance, if there is a set of tasks of different priorities requesting the same resources, it could easily result in priority inversion: a low-priority task pre-empted by a higher-priority one is blocking a resource and, thus, the execution of all those that wait for the same resource. In this case the waiting time cannot be bounded. The situation is depicted in Figure 2.6.

One of the attempts to cure this inherent problem uses priority inheritance; see Figure 2.7. However, the system is now behaving differently than specified, namely a task is given a different priority than originally assigned, which disturbs the global relationships, especially if the resource is allocated for a longer time. The higher-priority tasks are still delayed by the lower, although for a shorter time. Further, a track must be kept of which resources are allocated by which tasks.

Another widely used, but only slightly better policy is round-robin scheduling, where each of the ready tasks is assigned a time slice. Needless to say, the policy is not feasible unless the temporal circumstances of all competing tasks are more or less the same.

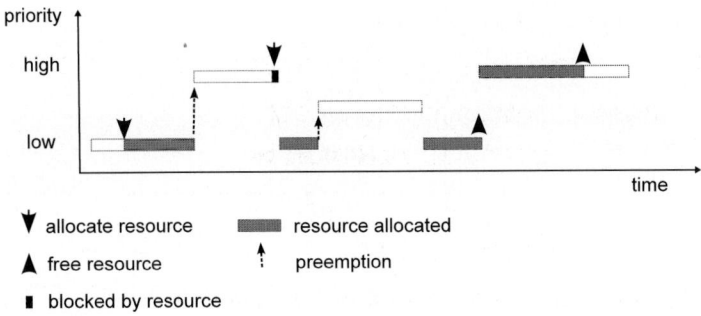

Fig. 2.6. Priority inversion

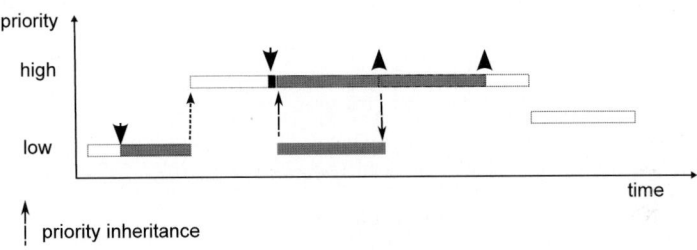

Fig. 2.7. Priority inheritance

Rate-Monotonic Priority Assignment

There are other policies to assign priorities to tasks which are more adequate to solve the scheduling problem. Well-known and particularly suitable for periodic tasks is rate-monotonic scheduling. It follows the simple rule that priorities are assigned according to increasing periods of periodic tasks — the shorter the period the higher the priority. The policy is feasible, and it is possible to carry out schedulability tests.

In their widely recognised publication [80], in 1973 Liu and Layland considered periodic task sets whose relative deadlines equal their periods. All n tasks were assumed to start simultaneously at $t=0$ and to be independent, *i.e.*, there exist no precedence relations or shared resources. Under these assumptions they proved that rate-monotonic priority assignment yields a feasible schedule *if*

$$U \leq n(2^{1/n} - 1) \; , \tag{2.1}$$

where U is the processor utilisation

$$U = \sum_{i=1}^{n} \frac{C_i}{T_i} \; . \tag{2.2}$$

The utilisation of the processor by n rate-monotonically scheduled tasks is shown in Table 2.1.

Table 2.1. Processor utilisation of systems with N tasks

N	Utilisation bound
1	100%
2	82.8%
3	78.0%
5	74.3%
10	71.8%

As an illustration, a Gantt chart of the feasible schedule of three tasks with their priorities assigned rate-monotonically is shown in Figure 2.8. The overall utilisation is 0.72, with the maximum for three tasks (from Table 2.1) being 0.78.

	Period	Exec. time	Utilisation
Task 1	15	4	0.27
Task 2	20	5	0.25
Task 3	25	5	0.20

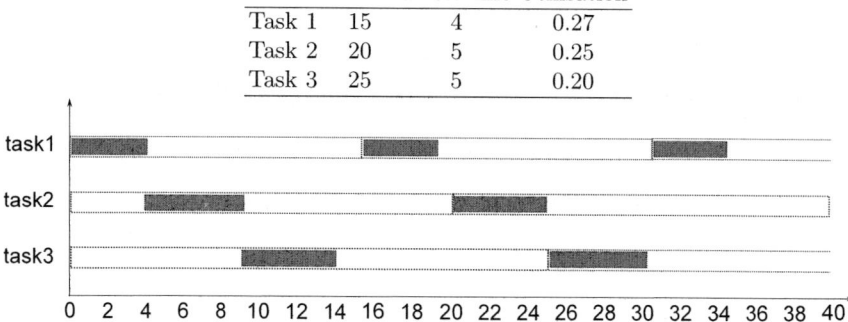

Fig. 2.8. Example of feasible rate-monotonic scheduling

In the next example in Figure 2.9, the attempted utilisation is 85%. As a result, Task 3 only receives four units of service time in its first period, is then pre-empted by Task 1, followed by Task 2, and misses its deadline.

For a large number of tasks, Equation 2.1 returns the feasibility bound

$$\lim_{n\to\infty} n(2^{1/n} - 1) = \ln 2 \cong 0.69 \ .$$

Thus, it is proven that the policy provides an adequate and feasible schedule for any set of ready tasks as long as the utilisation of the processor is below 0.69. Renouncing 30% of processor performance and using rate-monotonicity, it can be guaranteed that the schedule is always feasible.

The feasibility bound of Equation 2.1 is sufficient but not necessary, *i.e.*, it is possible that a task set will meet its deadlines under rate-monotonic scheduling although the condition is violated. Bini *et al.* [8] proved that a periodic task set is feasible under rate-monotonic scheduling as long as

$$\prod_{i=1}^{n} \left(\frac{C_i}{T_i} + 1 \right) \leq 2 \ .$$

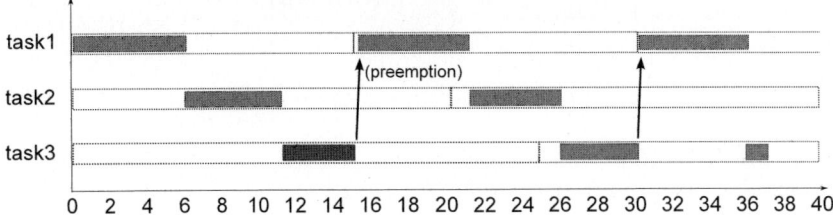

	Period	Exec. time	Utilisation
Task 1	15	6	0.40
Task 2	20	5	0.25
Task 3	25	5	0.20

Fig. 2.9. Example of overload with rate-monotonic scheduling

This schedulability test, which is known as the Hyperbolic Bound, allows higher processor utilisation. It takes into account that the schedulability bound improves if the periods have harmonic relations.

The rate-monotonic scheduling policy relates to periodic tasks. With the assumption that non-periodic tasks can only be invoked once within a certain period of time, in the worst case sporadic tasks become periodic; thus, rate-monotonic scheduling can also be used for them.

2.2.2 Deadline-driven Scheduling

Following the ultimate requirement for real-time systems, *viz.*, that all tasks meet their deadlines, leads to problem-oriented deadline-driven scheduling. The artificial priorities of tasks are no more relevant. In some adequate scheduling algorithms, priorities are only used in order to handle system overloads; in this case, priorities may indicate which tasks must be retained under any circumstances, and which ones may be gracefully degraded or renounced.

There are several deadline-based algorithms, actually rate-monotonic scheduling as described above being one of them. Some observe deadlines themselves, and others schedule tasks according to slack or margin (the difference between the time to the deadline and the remaining execution time of a task).

One of the most suitable scheduling algorithms is the *earliest deadline first scheduling algorithm* [80]. Tasks are scheduled in increasing order of their deadlines, see Figure 2.10. It was shown that this algorithm is feasible for tasks running on single processor systems; with the so-called throw-forward extension it is also feasible on homogeneous multiprocessor systems. However, this extension leads to more pre-emptions, is more complex and, thus, less practical. To be able to employ the strategy earliest deadline first, it is best to structure real-time computer systems as single processors. The idea can be extended to distributed systems, by structuring them as sets of interconnected uni-processors, each of which dedicated to control a part of an external environment. This is actually not a very critical limitation because of the nature of application tasks occurring in control.

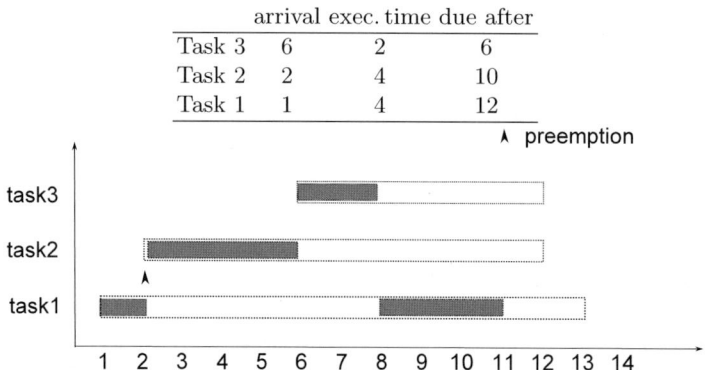

	arrival	exec. time	due after
Task 3	6	2	6
Task 2	2	4	10
Task 1	1	4	12

Fig. 2.10. An example of earliest deadline first scheduling

To handle the case of overloads, many researchers are considering load sharing schemes which migrate tasks between the nodes of distributed systems. In industrial process control environments, however, such schemes are generally not applicable, because only computing tasks can be migrated. In contrast to this, control tasks are highly input/output bound, and the permanent wiring of the peripherals to certain nodes makes load sharing impossible. Therefore, the earliest deadline first scheduling algorithm is to be applied independently from considerations of the overall system load on each node in a distributed system. The implementation of this scheme is facilitated by the fact that, typically, industrial process control systems are already designed in the form of co-operating, possibly heterogeneous, single processor systems, even though the processors' operating systems do not (yet) schedule by deadline.

Another scheduling policy based on deadlines, *least laxity first*, schedules tasks according to their slack. It was shown that it is feasible for both single- and multiprocessor systems. By the algorithm, the task with the least time reserve is executed first; see Figure 2.11. As a consequence, its reserve is maintained (accumulated execution time and the distance to the deadline are both running at the same speed), but the reserves of other, waiting tasks vanish. Thus, the reserve of one or more tasks being the next smallest reserve eventually reaches the executing task's reserve. From this moment on, the processor must be shared among the two or more tasks in round-robin fashion. Now, their reserves are not maintained anymore, since their execution is shared, but they expire more slowly than the ready tasks not executing. Sooner or later, the reserves of other waiting tasks will reach theirs, so that these tasks also join the pool for switched execution. Obviously, the overhead for context-switching caused in this way is high and usually unacceptable. For that reason, the least laxity first algorithm is mainly of theoretical interest only.

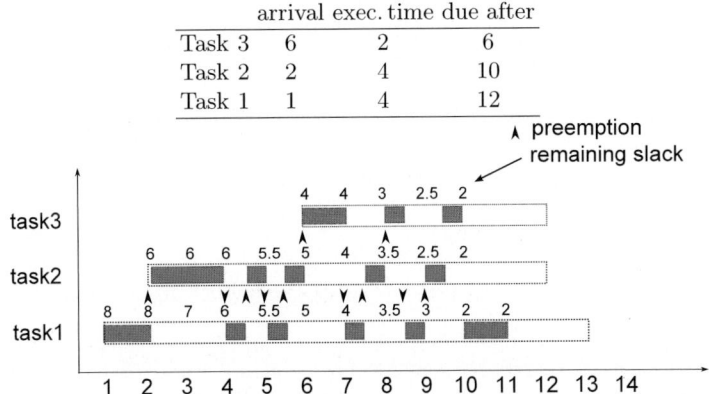

	arrival	exec. time	due after
Task 3	6	2	6
Task 2	2	4	10
Task 1	1	4	12

Fig. 2.11. An example of least laxity first scheduling

In contrast to least laxity first, the earliest deadline first algorithm does not require context-switches unless a new task with an earlier deadline arrives or an executing task terminates. In fact, if the number of pre-emptions enforced by a scheduling procedure is considered as a selection criterion, the earliest deadline first algorithm is the optimum one. Even when tasks arrive dynamically, this policy maintains its properties and generates optimum pre-emptive schedules [74].

This scheduling strategy establishes the direction in which an appropriate architecture should be developed. The objective ought to be to maintain, as much as possible, a strictly sequential execution of task sets. The processor(s) need(s) to be relieved of frequent interruptions caused by external and internal events in the sequential program flow. These interruptions are counterproductive in the sense that they seldom result in immediate (re-) activation of tasks.

2.2.3 Sufficient Condition for Feasible Schedulability Under Earliest Deadline First

When deadlines and processing requirements of tasks are available *a priori*, the earliest deadline first algorithm schedules the ready tasks by their deadlines. For any time t, $0 \leq t < \infty$, and any task $T \in \mathcal{F}(t)$ with deadline $t_z > t$, from the set of ready tasks $\mathcal{F}(t)$ with n elements, let

$a(t) = t_z - t$ be its response time,
$l(t)$ the (residual) execution time required before completion, and
$s(t) = a(t) - l(t)$ its laxity (slack-time, margin, time reserve).

Then, necessary and sufficient conditions that the task set $\mathcal{F}(t)$, indexed according to increasing response times of its n elements, can be carried through meeting all deadlines for single processor systems, are

$$a_k \geq \sum_{i=1}^{k} l_i, \; k = 1, ..., n \qquad (2.3)$$

This necessary and sufficient condition determines whether a task set given at a certain instant can be executed within its specified deadlines. In words, it reads

> *If it holds for each task that its response time (the time until its due date) is greater then, or equal to, the sum of its (residual) execution time and the execution times of all tasks scheduled to be run before it, the schedule is feasible.*

In the ideal case, the earliest deadline first method guarantees one-at-a-time scheduling of tasks, modulo new task arrivals. Thus, unproductive context-switches are eliminated. Furthermore, and even more importantly, resource access conflicts and many concurrency problems among the competing tasks, such as deadlocks, do not occur and, hence, do not need to be handled. Unfortunately, such an ideal case is by its very nature unrealistic. In the following, more realistic conditions will be considered.

Earliest Deadline First Scheduling under Resource Constraints and Feasible Schedulability Conditions

The above considerations have revealed that the earliest deadline first algorithm is the best choice for use in a general multitasking environment, provided that the tasks are pre-emptable at any arbitrary point in time. Unfortunately, this pre-condition is not very realistic, since it is only fulfilled by pure computing tasks fully independent of one another. In general, however, tasks have resource requirements and, therefore, execute critical regions to lock peripherals and other resources for exclusive and uninterrupted access. Hence, the elaboration of a task consists of phases of unrestricted pre-emptability alternating with critical regions. While a task may be pre-empted in a critical region, the pre-emption may cause a number of problems and additional overheads, such as the possibility of deadlocks or the necessity for a further context-switch, when the pre-empting task tries to gain access to a locked resource. Therefore, to accommodate resource constraints, the earliest deadline first discipline is modified as follows:

> *Schedule tasks earliest deadline first, unless this calls for pre-emption of a task executing in a critical region. (In this case, wait until the task execution leaves the critical region.)*

Note that task precedence relations do not need to be addressed here, since the task set in consideration consists of ready tasks only, *i.e.*, the set consists of tasks whose activation conditions are fulfilled (and, thus, their predecessors have already terminated).

If all tasks competing for the allocation of the processor are pre-emptable, then the entire set is executed sequentially and earliest deadline first. Therefore, the (partial) non-pre-emptability of certain tasks can only cause a problem if a newly arrived task's deadline is shorter than that of the task running, which is at that same time executing in a critical region. Hence, all tasks with deadlines closer than that of the executing task, including the newly arrived one, have to wait for some time d, before the running task can be pre-empted. Practically it is not feasible to determine the exact value of d. The most appropriate upper bound for d is given by the maximum of the lengths of all non-pre-emptive regions of all tasks in a certain application, since this is the most general expression fully independent on the actual time and the amount of resource contention. With this, a sufficient condition, that allows one to determine a task set's feasible schedulability at any arbitrary time instant under the above algorithm, reads as follows:

If a newly arrived task has an earlier deadline than the executing one, which is just within a critical region, the schedule according to the above algorithm is feasible if:
(a) all tasks, which should precede the running one (Tj) according to the earliest deadline first policy, have their response times greater than, or equal to, the sums of (1) their execution times, (2) those of all tasks in the schedule before them, and (3) d, the time required to leave the critical region,

$$a_i(t) \geq d + \sum_{k=1}^{i} l_k(t), \ \ i = 1, ..., j-1 \qquad (2.4)$$

and

(b) all other tasks have their response times greater than, or equal to, the sum of their (residual) execution times and the execution times of all tasks scheduled to be run before them:

$$a_i(t) \geq \sum_{k=1}^{i} l_k(t), \ \ i = j, ..., n \qquad (2.5)$$

We observe that, if there are no critical regions, this condition reduces to the one holding for feasible schedulability under no resource constraints. When running tasks may not be pre-empted at all, a newly arrived task with an earlier deadline has to wait until the running task terminates. A sufficient condition for feasibly scheduling non-pre-emptive tasks follows as an easy corollary from the result mentioned above.

This modified earliest deadline first policy offers a practical way to schedule tasks predictably in the presence of resource constraints. The policy maintains most of the advantageous properties as listed in Section 2.2.4. In fact, the only property no longer exhibited is the attainability of maximum processor utilisation because of the pessimistic prediction of the delay in pre-empting the executing task. From the point of view of classical computer performance theory, this may be considered a serious drawback. For embedded real-time systems, however, it is not so relevant whether processor utilisation is optimum, as costs have to be seen in the framework of the controlled external process, and with regard to the latter's safety requirements. Taking the costs of a technical process and the possible damage into account which a processor overload may cause, the cost of a processor is of less importance. Moreover, while industrial production cost in general increases with time, the cost of computer hardware decreases. Hence, processor utilisation is not a suitable design criterion for embedded real-time systems. Lower processor utilisation is, thus, a small price to be paid for the simplicity and the other advantageous properties of the scheduling method presented here, which yields high dependability and predictability of system behaviour.

Avoiding Context-Switches Without Violating Feasible Schedulability

So far we have neglected the overhead costs due to context-switching. Now we demonstrate how some of the context-switches can be avoided, and how the cost of the remaining switches can be taken into account. Let us assume that the time required to prepare a task's execution and to load its initial context into the processor registers as well as to remove the task from the processor after the task has terminated normally is already included in the maximum execution time specified for the task. This assumption is realistic, since these context-changes have to be carried out under any circumstances and are not caused by pre-emptions. Thus, we only have to account for the time required for a pre-emptive switch. Without loss of generality, we further assume that this time is the same for all tasks, and denote it by u, and that to either save or restore a task takes $u/2$.

The following modified form of the earliest deadline first task scheduling algorithm not only takes resource constraints into account, but also avoids superfluous pre-emptions:

> *Always assign the ready task with the earliest deadline to the processor, unless new tasks with earlier deadlines than the deadline of the currently running task arrive. If the laxities of the newly arrived tasks allow for feasible non-pre-emptive scheduling, then continue executing the running task. Otherwise, pre-empt the task as soon as its critical region permits, and allocate the processor to the task with the earliest deadline.*

Let us examine what occurs, when a pre-emption is required:

> *If a newly arrived task has an earlier deadline than the executing one, which is just within a critical region, the schedule according to the above algorithm is feasible, if:*

(a) all tasks, which should precede the running one according to the earliest deadline first policy, have their response times greater than, or equal to, the sums of (1) their execution times, (2) those of all tasks in the schedule before them, (3) the time d required to leave the critical region, and (4) a half of the switching time u/2,

$$a_i(t) \geq d + u/2 + \sum_{k=1}^{i} l_k(t), \ i = 1, ..., j - 1 \qquad (2.6)$$

and

(b) all other tasks have their response times greater than, or equal to, the sum of their (residual) execution times, the execution times of all tasks scheduled to be run before them, and the full switching time u:

$$a_i(t) \geq u + \sum_{k=1}^{i} l_k(t), \ i = j, ..., n \qquad (2.7)$$

Note that only the status-saving part of the context-switch needs to be considered in the feasibility check of tasks scheduled to be run before the pre-empted one, since they already contain their start-up time. For all other tasks, the total context-switching time u has to be taken into account. If there is no pre-emption, the feasibility check for non-pre-emptable scheduling applies.

2.2.4 Implications of Employing Earliest Deadline First Scheduling

To summarise, earliest deadline first task scheduling on a single processor system has the following advantages:

- Scheduling on the basis of task deadlines is problem-oriented,
- It allows one to formulate tasks and to extend and modify existing software without the knowledge of the global task system,
- All tasks can be treated by a common scheduling strategy, regardless of whether they are sporadically or periodically activated, or have any precedence relations,
- Feasibility,
- Upon a dynamic arrival of a ready task, the task's response time can be guaranteed (or a future overload can be detected),
- The cost of running the algorithm is almost negligible (its complexity is linear in the number of tasks in the ready queue),
- Ease of implementation,
- The cost of checking a task set's feasible schedulability is almost negligible (again, linear in the number of tasks in the ready queue), and the check itself is trivial, *i.e.*, the operating system is enabled to supervise if the fundamental timeliness condition is met,

- Facilitation of early overload detection and handling by dynamic load adaptation, thus allowing system performance to degrade gracefully,
- Achieving the minimum number of task pre-emptions required to execute a feasible schedule,
- Achieving maximum processor utilisation while maintaining feasible schedulability of a task set,
- It is essentially non-pre-emptive, *i.e.*, task pre-emptions may only be caused when dormant tasks are activated or suspended ones are resumed,
- The sequence of task executions is determined at the instants of task (re-) activations and remains constant afterwards, *i.e.*, when a new task turns ready, the order among the others remains constant,
- The order of task processing is essentially sequential,
- Resource access conflicts and deadlocks are inherently prevented,
- Unproductive overhead is inherently minimised,
- The priority inversion problem, which has received much attention, does not arise at all, and
- Pre-emptable and (partially) non-pre-emptable tasks can be scheduled in a common way.

Apart from these implications, there is another important one to be considered here: earliest deadline first scheduling directly supports reconfiguration of a system as an adequate measure of fault tolerance. If any change to the tasks occurs (either graceful degradation and thus smaller execution time, or complete omission in order to provide more time to other ones), no additional change other than the tasks' parameters is necessary. The scheduling system automatically adapts to the new situation.

2.2.5 Rate Monotonic vs Earliest Deadline First Scheduling

In this section, the two most appreciated scheduling policies, *viz.*, rate-monotonic and the earliest deadline first, are compared. Although the earliest deadline first algorithm has various interesting advantages compared to rate-monotonic scheduling, the latter is applied more often in practice. The comparison provided in this section shows that merely implementation aspects are in favour of the rate-monotonic approach in all other aspects earliest deadline first provides better or at least equal performance. A considerable fraction of the facts listed below are taken from [12], where the interested reader can find further details.

Implementation Complexity

A common misconception is that rate-monotonic scheduling causes far less implementation complexity than earliest deadline first scheduling. Both algorithms require one to administrate a list of activated tasks. A rate-monotonic scheduler selects the activated task with the highest priority, whereas the earliest deadline first policy requires one to look for the task with the closest absolute deadline. The computational effort to order the list of activated tasks by priorities or by absolute deadlines differs only insignificantly, due to the higher word length required for time values. Only the fact that the earliest deadline first algorithm computes a new absolute deadline at each task activation causes an observable difference in computing complexity. This, however, can be neglected as long as the time values are handled as binary numbers and not in a representation conforming to Universal Time Co-ordinated (UTC). Since most real-time operating systems are based on priority-based scheduling, the earliest deadline first algorithm is sometimes implemented by mapping deadlines to priorities. However, this workaround implementing earliest deadline first is neither easy nor efficient.

Runtime Overhead

The slightly higher implementation complexity of the earliest deadline first algorithm results in a little greater need for execution time of the operating system kernel functions. Nevertheless, this algorithm causes less runtime overhead. This is because the average number of pre-emptions is significantly higher for rate-monotonic than for earliest deadline first scheduling. Hence, the runtime overhead due to context-switching is considerably higher. Figure 2.12 exemplifies how rate-monotonic scheduling leads to unnecessary task pre-emptions. The simulation results presented in [12] show that the ratio of the average number of pre-emptions induced by rate-monotonic scheduling to those induced by earliest deadline first scheduling is higher the higher the number of tasks or the system load. This reflects the fact that a higher number of tasks as well as longer execution times increase the chance that a higher-priority task is activated during a task's execution.

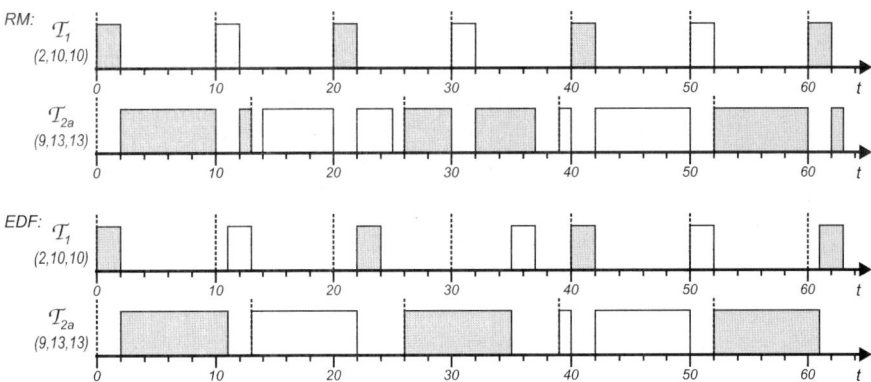

Fig. 2.12. Scenario in which rate-monotonic scheduling induces dispensable task pre-emptions compared to earliest deadline first scheduling. Succeeding instances of the tasks are displayed in different colours

Schedulability

The computational effort to check feasibility depends strongly on the temporal characteristics of the task set analysed, *i.e.*, the application software. All schedulability conditions presented in the previous sections for both rate-monotonic and deadline-driven scheduling are based on the assumption that tasks are pre-emptable at any point of execution. Under this assumption, the optimality theorem of Dertouzos [26] applies:

> *The earliest deadline first algorithm is optimum in the sense that, if there exists any algorithm that can build a valid (feasible) schedule on a single processor, then the earliest deadline first algorithm also builds a valid (feasible) schedule.*

According to that, a task set proven feasible under rate-monotonic scheduling is also feasible under earliest deadline first scheduling. This, however, does not hold the other way. Figure 2.13 illustrates the higher schedulability bound of earliest deadline first scheduling for an example.

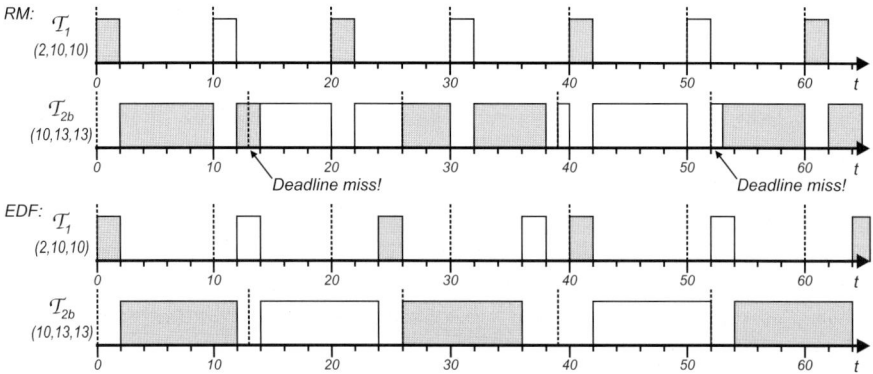

Fig. 2.13. Scenario in which the earliest deadline first algorithm creates a feasible schedule, but the rate-monotonic algorithm does not. Succeeding instances of the tasks are displayed in different colours

If tasks contain non-interruptible program parts, the computational complexity of both scheduling algorithms increases significantly. Most methods proposed in the literature compensate for this by adding a delay to each task's execution time. This delay results from the duration of the non-interruptible program segments of all other tasks that can fall within a task's execution. In rate-monotonic scheduling, a task can only be delayed by a task with a lower priority if this lower-priority task is running and in a critical section at the activation instant. Hence, the approach of assuming a delay is less pessimistic for tasks with high priorities. On the other hand, tasks with long periods are pre-empted more often in rate-monotonic scheduling than in earliest deadline first scheduling. This causes more pessimistic delay assumptions for tasks with lower priorities. The characteristics of earliest deadline first scheduling can also be used to obtain more precise delay assumptions.

In summary, the feasibility analysis of task sets with critical sections has different characteristics for the two scheduling policies. Earliest deadline first has the higher schedulability bound in general, but the specific characteristics of rate-monotonic scheduling might be beneficial to analyse feasibility in certain applications.

Resource Sharing

Mutually exclusive access to shared resources requires one to take the priority inversion phenomenon into account and to apply mechanisms that exclude deadlocks. Although most concurrency control protocols, like, *e.g.*, the Priority Inheritance Protocol (PIP) and the Priority Ceiling Protocol (PCP) [100], were originally devised to solve this problem under rate-monotonic scheduling, their underlying principles are applicable to both scheduling algorithms [106].

Overload Behaviour

Under rate-monotonic scheduling a permanent overload causes complete blocking of low-priority tasks, whereas in case of earliest deadline first scheduling tasks behave as they were activated at a lower rate [12]. Both characteristics can be useful in certain applications. Rate-monotonic scheduling has the advantage that an overrun of a task cannot cause tasks with higher priorities to miss their deadlines, while an overrun under earliest deadline first scheduling can affect all tasks. Since the temporal behaviour changes significantly in overload situations, however, highly safety-critical applications require one to exclude them by design. The acceptance of overload situations would unnecessarily increase the effort for proving safety dramatically, since all possible overload scenarios would have to be analysed.

Jitter and Latency

The absolute response jitter, *i.e.*, the difference between the actual maximum and minimum response time of a task, is an important measure of real-time performance. Since high jitter can cause instability of control loops, low jitter is desirable. Rate-monotonic and earliest deadline first scheduling have different jitter characteristics. As the simulation results in [12] show, the rate-monotonic algorithm realises low jitter for high-priority tasks at the price of high jitter for low-priority tasks. Earliest deadline first scheduling treats tasks more evenly. Compared to rate-monotonic scheduling, tasks with short periods have slightly higher jitter, but tasks with long periods have considerably smaller jitter. Earliest deadline first scheduling can also provide very low jitter for certain tasks if the results of a task are output by a separate output task with a very short relative deadline. Since this constrains feasibility, the small jitter of that particular task must be realised at the expense for longer relative deadlines of the former tasks — just like under rate-monotonic scheduling. Hence, rate-monotonic scheduling does not clearly outperform the earliest deadline first algorithm in terms of jitter.

Input-output latency is another measure of real-time performance, that is important, *e.g.*, in terms of control stability. Cervin proved that input-output latency under earliest deadline first scheduling is always less than or equal to that of rate-monotonic scheduling [13].

Summary of the Comparison

The argument that rate-monotonic scheduling causes less implementation complexity holds for work-around implementations on top of priority-based real-time operating systems, only. Rate-monotonic scheduling provides the advantages of timely execution in overload situations and less jitter just for high-priority tasks, whereas earliest deadline first scheduling treats all tasks evenly. The major advantage of earliest deadline first scheduling is its higher schedulability bound. This enables higher exploitation of the computing capacity and more efficient implementation of periodic or aperiodic servers. Since most resource reservation strategies are based on mechanisms similar to these servers, they also provide better performance under earliest deadline first scheduling [12].

2.3 Synchronisation Between Tasks

A task being executed may require certain resources that can be available or not. Further, there may be precedence relations among tasks that require a task's execution to be postponed to the time when these conditions are fulfilled. The mechanism allowing for these and further related features is *synchronisation*.

It may be noted at the beginning that in some cases, *e.g.*, providing for mutual exclusion of tasks, synchronisation problems arise as soon as dynamic and asynchronous pre-emptive tasking is involved. If systems can be implemented in a static manner, *e.g.*, using the cyclic executive approach, or can be statically scheduled in any other way, no synchronisation problems can arise. Even in the case of dynamic scheduling, by selecting an appropriate scheduling paradigm (*e.g.*, earliest deadline first), and by renouncing pre-emptions, tasks are not forced to check out at arbitrary instants. Thus, the critical sections are always executed in a non-interruptible way and there is no need for other mechanisms providing for that.

Unfortunately, this is not always possible; as a rule, dynamic asynchronous systems require pre-emptions to guarantee asserted response times of sporadically appearing tasks and, thus, some sort of synchronisation mechanisms are also needed for mutual exclusion and similar features.

As can be seen in the task state transition scheme in Figure 2.3, the executing task is suspended from the "running" state into a state where it waits for resumption. On any state change of synchronisation primitives a mechanism is checking whether a task from the "suspended" state can be resumed and, in this case, carries out the state change.

When in the states "suspended", "ready", or "running", tasks have already been activated and their response times $a(t)$ are running out. There is, however, a difference between the "suspended" and the other two states: as in the cases of feasible scheduling policies the scheduler is keeping track that deadlines are met; it cannot control the waiting of tasks in the "suspended" state until they are resumed to the "ready" state. Meanwhile, deadlines may already have passed, or laxities $s(t)$ may have become smaller than remaining execution times $l(t)$, which means that deadlines will be missed.

It is a severe problem that waiting in the "suspended" state is non-deterministic and non-predictable. For this reason, and to allow one to meet suspended tasks' deadlines, it is necessary to bound the waiting time in this state: after a task has waited for a certain predefined time, it is released from the waiting state and a predefined action is taken to resolve the situation. In the worst case, the task has failed and is aborted to the state "dormant".

Mutual Exclusion

When different tasks require access to the same resources (peripherals, common variables, *etc.*), a mechanism must be employed to prevent simultaneous actions resulting in erroneous states of the shared resources. A *critical section* is part of a task's program for which the task requires exclusive access to a shared resource. The mechanism preventing other tasks from simultaneously accessing it is *mutual exclusion.*

A typical example is a test-and-set operation on a flag: with its first instruction the state of a Boolean variable is read, with the second one tested, and with the third one set, if the test yielded that the flag marks free. Clearly, if two tasks would execute the same instructions simultaneously, they would both find the flag signaling free and would set it.

Fully consistent methods to implement mutual exclusion are based on hardware. In Motorola's 68K microprocessor family, for instance, the instruction Test_And_Set (TAS) performs this action in a non-interruptible manner. Hence, double access is physically prevented. If this is not possible, mutual exclusion must be implemented with programmed synchronisation mechanisms. There is a number of them which are more or less suitable for application in embedded real-time systems. A competent and in-depth description can be found in [6]. Below, some of the most common concepts based on shared variables will be sketched briefly.

2.3.1 Busy Waiting

The simplest synchronisation mechanism consists of waiting in an idle loop while the condition for entering a critical section is not fulfilled. This procedure is also called "spinning" and the flag variable "spin lock". Before entering a critical section, a task tests whether a flag (initially set to "free") indicates that no other task is in its critical section. If yes, it locks the flag and enters. If not, it waits until some other task unlocks it.

If there are two tasks, each with a critical section, two flags are needed in order to avoid simultaneity problems: before entering its critical section, each task first sets its own flag to indicate its intention and then checks the other task's flag. If the latter does not indicate free, the task spins in a loop; otherwise the task enters its critical section and sets its own flag to free on exiting; see Figure 2.14.

```
Task1:                              Task2:
      flag1:=locked;                      flag2:=locked;
      while flag2=locked do               while flag1=locked do
         nothing                             nothing
      end do;                             end do;
      run critical section;               run critical section;
      flag1:=free;                        flag2:=free;
      ...                                 ...
```

Fig. 2.14. Busy waiting, first example

An alternative procedure is first to check the other task's flag and eventually wait and then to seize its own when entering the critical section; see Figure 2.15.

```
Task1:                              Task2:
      while flag2=locked do               while flag1=locked do
         nothing                             nothing
      end do;                             end do;
      flag1:=locked;                      flag2:=locked;
      run critical section;               run critical section;
      flag1:=free;                        flag2:=free;
      ...                                 ...
```

Fig. 2.15. Busy waiting, second example

Busy waiting is inefficient, since during waiting the processor is occupied by running a useless loop. Also, it is prone to *livelock*: let us consider a situation in Figure 2.14 when a task sets its flag, then unsuccessfully checks the other task's flag, and goes into a waiting loop. The other task does exactly the same at the same moment; thus, both tasks wait running in idle loops. Further, since the mechanism is performed by the tasks themselves, it may be pre-empted during its execution. Such a situation may, for example, happen in the second solution of Figure 2.15, when both flags indicate free in the beginning. The first task checks the flag, is pre-empted, then the second one does the same, and is pre-empted too. Now the first task locks the flag and enters the critical region, is pre-empted, and now the second task does exactly the same. Thus, they are both in the critical sections.

2.3.2 Semaphores

Semaphores, introduced by Dijkstra, are implemented by two procedures, P and V; for clarity, however, we shall use the alternative names:

WAIT(S) : if $S > 0$ then $S = S - 1$ else suspend the running task;

SIGNAL(S) : if no task is suspended on S then $S = S+1$ else resume one task.

Generally, a semaphore is a non-negative integer value. In this case it is called *counting* semaphore. If the values are 0 and 1 only, it is known as *binary* semaphore.

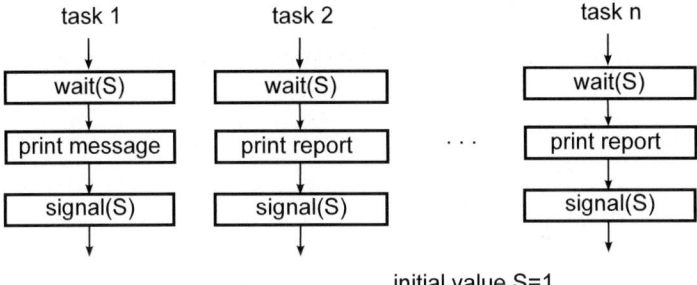

Fig. 2.16. Printer server, first example

Mutual exclusion is achieved if each task performs WAIT(S) before entering its critical section, and SIGNAL(S) on leaving it. The initial value of the semaphore S is 1. The first task which succeeds in seizing the semaphore wins, the other(s) have to wait. If there are more resources available, S is initialised by their number. Tasks each obtain an instance of the resource, until there are none left. An example of a printer server is shown in Figure 2.16. If more printers are in a system, S can be initialised by n; the printer driver then assigns printing jobs to different printers as long as they are available. Another possibility to implement a printer server is shown in Figure 2.17.

If the procedures WAIT and SIGNAL are executed without being interrupted, there can be no simultaneity problems accessing synchronisers. In other words, tasks that execute the two procedures may not be pre-empted.

Employing semaphores there is the danger of causing deadlocks. Let us assume that a task A is waiting for S1 before it can release (signal) S2. Task B,

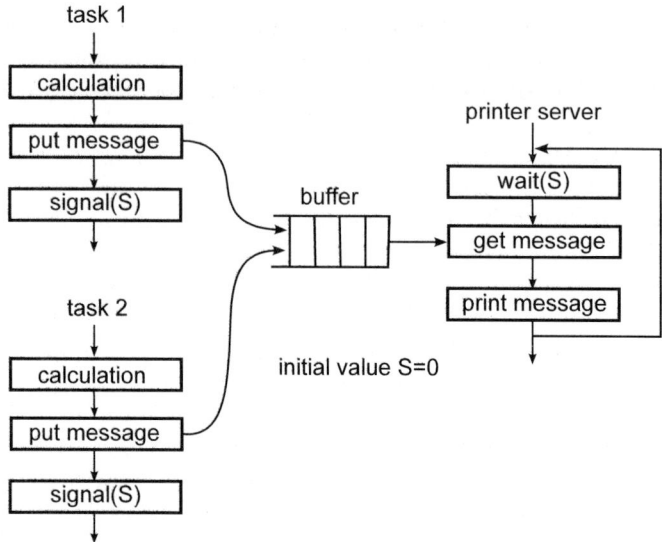

Fig. 2.17. Printer server, second example

on the other hand, is waiting for S2, which prevents it from releasing S1. The situation is known as *deadlock*, since none of the tasks involved can proceed and both are, thus, stuck indefinitely long in this situation. Another danger is *starvation*. Due to an unfortunate sequence it may happen that a task can never gain access to the semaphore, or may wait for it non-predictably and unacceptably long.

2.3.3 Bolts

An interesting generalisation of semaphores are bolts. They were introduced in the process automation programming language PEARL [29], and are useful for situations when some tasks only wish to read resources (*e.g.*, common variables or peripherals) whereas the others need to change their states. Thus, the reading tasks could be allowed shared access, but only if the states are not changed at the same time by the writing tasks. The latter, naturally, need exclusive access to the shared resources.

Bolts may acquire the values "locked" (*bolt* = −1), "free" (or lockable, *bolt* = 0), or "shared" (or non-lockable, 0 < *bolt* < *range*). There are four operations on a bolt:

RESERVE: if the state of the bolt is "free" (lockable), it is locked; if not, the executing task is suspended in a RESERVE queue.

FREE: the bolt becomes "free". If there are any tasks waiting in its RESERVE queue, they are put to the ready state and the one with the highest priority locks the bolt. If the RESERVE queue is empty, tasks from the ENTER queue are released.

ENTER: if the bolt is locked or there are tasks in the RESERVE queue, the executing task is suspended in the ENTER queue. If not, the task puts the bolt in the "shared" state (if it is not already) by incrementing its value.

LEAVE: the value of the bolt is decremented. If it reaches 0 (the task was the only one in the shared critical section), the bolt becomes "free" and the tasks in the RESERVE queue are released, if any.

2.3.4 Monitors

A monitor is a structured synchronisation construct with a higher level of abstraction that is supported by programming languages rather than operating systems. It has been introduced by Hoare in 1974. His purpose was to relieve programmers from explicit handling of semaphores, which is not easily understood and error-prone if, *e.g.*, any of the semaphore operations is omitted or mistaken. Being a structured construct, that cannot happen with monitors. There are two basic ideas monitors are built on:

- Monolithic monitors centralise critical functions; they run in an uninterruptible mode, can access protected memory areas, and execute privileged instructions;
- They structure the critical synchronisation functions and put them under system supervision.

Monitors encapsulate shared data with procedures that incorporate the critical sections. Their execution cannot be pre-empted. A monitor has a single lock, and a task that wants to execute monitor functions must acquire it exclusively; it is then called *active task*. That is the only task that can activate the monitor's functions.

In a monitor there is any number of *event-* or *condition variables queues*. A condition variable is not actually a variable, but a condition that is handled with two monitor methods associated with that queue, typically called *wait* and *signal*. In these queues tasks wait for certain conditions (*e.g.*, availability of a resource) to be fulfilled. An active task may release the monitor lock and enter a queue by executing the corresponding *wait* command. Different to semaphores, the monitor *wait* always blocks the executing task. It may also unblock a task that is waiting in a queue by sending a *signal* to that queue. If no task is waiting in this queue, the *signal* has no effect. To enable these actions, there are three further types of queues in a monitor; see Figure 2.18:

- *Entry queue:* when a task attempts to enter the monitor (to access a monitor method), it is put into this queue;
- *Signaler queue:* when a task performs a signal method, it is placed into this queue;
- *Ready queue:* a task is put into this queue, when it is removed from one of the condition variable queues.

The entry, signaler, and ready queues are called *monitor queues*; tasks are waiting in these queues to become active, *i.e.*, to acquire the monitor lock. The functions and priorities of the monitor queues define the nature of a monitor. They are not all necessary, *e.g.*, the signaler queue can be replaced by a certain behaviour of a signaling task, *viz.*, by immediately leaving the monitor.

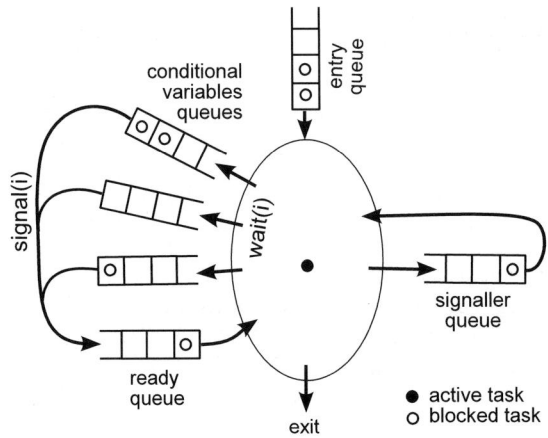

Fig. 2.18. Queues within a monitor

2.3.5 Rendezvous

Semaphores and monitors have the drawback of being centralised solutions: first, this presents a central point of failure and, second, embedded systems are by their nature becoming more and more distributed. In such systems, message-based synchronisation mechanisms are preferable. In multitasking environments, executing tasks run asynchronously. However, tasks may check certain conditions and, when fulfilled, decide whether to continue code execution or to wait further. Such conditions can be message arrivals from other tasks.

The idea of message-based synchronisation is that a task arriving at a certain place in a program first, sends a message to its partners and waits. This is called a rendezvous, was proposed by Hoare, and implemented in the language Ada. Its procedure is the following: a task Q is providing a shared resource to tasks P_i. When, *e.g.*, the task P_1 arrives to a place where this resource is needed, it waits until the task Q arrives to the place where it can offer the service. Then the part of Q is executed, and the tasks continue their execution separately. If the task Q arrives at the rendezvous first, it waits until one of the tasks P_i requires the service. The basic properties of this synchronisation concept are that waiting of the two tasks is symmetrical, that one transaction is performed during the rendezvous, that the other tasks P_i are mutually excluded, and that the tasks communicate by messages.

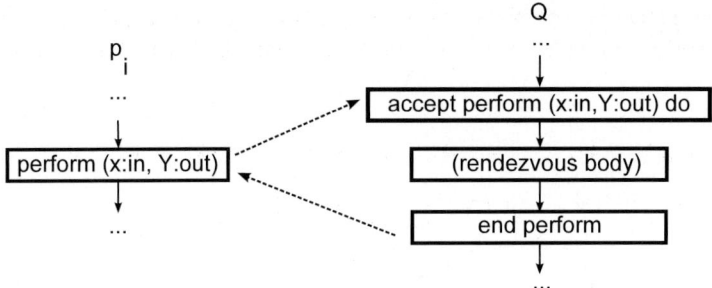

Fig. 2.19. Rendezvous operation and syntax

The concept is implemented by the *accept* statement which is syntactically similar to a declaration of a procedure (see Figure 2.19); it also includes input and output parameters. If the task P_i arrives at the place where the common code should be executed, it waits until Q reaches the *accept* statement. Alternatively, if Q first comes to the *accept* statement, it waits until one of the tasks P_i requires it. During execution of the accept part (between *do* and *end* in Ada) the calling task is blocked, and thus the *accept* acts as critical section.

In Figure 2.20 the problem of shared resources is solved using the rendezvous principle. If two tasks wish to use the same resource as process output, both can request it sending a message to the resource server. The message received first allocates it to its task. The resource is freed when the rendezvous is executed, and the two tasks that ran in common through the rendezvous are separated again.

2.3.6 Bounding Waiting Times in Synchronisation

When a task is suspended from execution and put into a waiting queue, it has no opportunity to follow its temporal circumstances like remaining slack time or time to its deadline. Also, it is out of reach of the scheduler and other operating system kernel mechanisms that could monitor the task and take care that it meets the temporal requirements imposed on it. The situation can become critical when a task is waiting too long for certain conditions to allow for its resumption of execution.

Therefore, it is necessary to introduce certain mechanisms providing means to tackle the above problem. This is much easier if the synchronisation constructs are structured. In this case, the time for waiting in a queue of suspended tasks can be bounded to certain predefined values. If exceeded, programmed actions can be carried through, followed by certain tasking operations like termination, activation, or abort. An example of shared objects

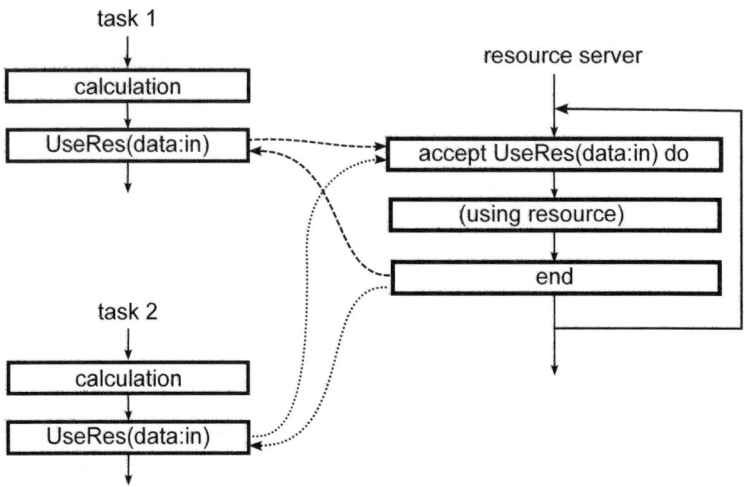

Fig. 2.20. Using rendezvous for exclusive use of an outbound resource

protected by implicit bolts is presented below. Accesses to them are enclosed in LOCK statements. The syntax is problem-oriented and allows one to:

- Express resource requests explicitly,
- Specify limits for waiting times and alternative reactions,
- Supervise the execution times of critical regions,
- Release resources as soon as possible, and
- Implement deadlock prevention schemes.

A syntax for structured synchronisation with bounded waiting times is shown in Figure 2.21. The statements between the PERFORM and UNLOCK clauses represent a critical region which is managed by synchronisation objects declared in the list. They are locked in an exclusive or shared manner (certain objects may only be locked exclusively, *e.g.*, semaphores). Pre-emption of execution within the critical region may be prevented. To bound waiting for synchronisers, a TIMEOUT clause may be defined. In the case of overrun, the OUTTIME statement string is executed to deal with the unsuccessful synchronisation. When the UNLOCK is reached, all synchronisers used are unlocked. With the QUIT statement it is possible to leave a critical region prematurely, unlocking all locked synchronisers. Within the region it is also possible to unlock a certain subset of them explicitly if they are not relevant any more.

```
structured-synchronisation-statement::=
    LOCK synchronisation-clause-list
        [NONPREEMPTIVELY]
        [timeout-clause] [exectime-clause]
    PERFORM statement-string UNLOCK;

synchronisation-clause-list::=
    synchronisation-clause {,synchronisation-clause}

synchronisation-clause::=
    EXCLUSIVE(synchronisation-object-list) |
    SHARED(synchronisation-object-list)

synchronisation-object-list::=
    synchronisation-object {,synchronisation-object}

synchronisation-object::=
    semaphore | bolt | shared-variable | ...

timeout-clause::=
    TIMEOUT
    {AFTER duration-expression | AT clock-expression}
    OUTTIME statement-string FIN

exectime-clause::=
    EXECTIMEBOUND duration-expression

quit-statement;;=
    QUIT;

unlock-statement::=
    UNLOCK synchronisation-object-list;
```

Fig. 2.21. Syntax of structured synchronisation construct

3

Hardware and System Architectures

Hardware architectures and their implementations provide the lowest level of control system design. Regarding the microscopic predictability approach [105], the higher layers of real-time systems are only predictable if the lower ones are. For that reason, extreme care ought to be taken when selecting approaches, guidelines, technologies and, particularly, off-the-shelf components in their design. The operation on the hardware layer must inherently be as safe as possible. It is not sufficient that hardware platforms are tested after being built; support for integrity and its validation must be considered an important guideline in their design. In spite of that, only little research dedicated to apt hardware solutions for embedded real-time applications can be found.

Generally, microprocessors and microcontrollers are becoming more and more complex. The market ratio between general-purpose and embedded applications a decade or two ago called, at that time, for more effort in designing the former. Thus, the main guideline in their design was to boost average performance, but not to improve their worst-case behaviour, often resulting in excessive complexity. The aim of these efforts was to achieve the best resource utilisation in order to accelerate processing in a carefully assessed most probable situation. The resulting processor complexity is unfortunate for the use of such microprocessors in embedded real-time applications. It renders almost impossible the assessment of the execution times for even the most simple sets of instructions, not to speak of verifying their correctness.

Today, in contrast, processor utilisation is an anachronism that should not be an issue any more at the current state of technology and well advanced techniques of distributed processing. Moreover, the ratio between embedded and desktop applications has turned around. For the year 2003 the following information could be found:

"In 2003, about 44 billion US$ worth of microprocessors were manufactured and sold. Although about half of that money was spent on CPUs used in desktop or laptop personal computers, those count for

only about 0.2% of all CPUs sold. The body count is dominated by 8-bit processors, followed by 16-bit and then 4-bit chips.
Source: World Semiconductor Trade Statistics Blue book, 2003.

In spite of the fact that this information is rather old and already outdated, it is realistic to presume that since then the situation has only become even more biased. In 2005, *e.g.*, only 2% of the money spent on microprocessors went to Pentium (Jim Turley, Silicon Insider, 2006.)

Considering this background, it is surprising that so little effort is devoted to the development of novel processor architectures explicitly dedicated to embedded applications. According to Moore's law – doubling the number of on-chip transistors every 18 months – we are getting enormous processing power which is now difficult to deploy. Instead of investing it in the overhead of sophisticated systems for enhancing its utilisation, at least in the case of processors for embedded applications, it should be used in a different way.

Following the pattern discussed in the section on general guidelines, the most important one is *simplicity*. It is impossible to ensure behavioral pre-dictability of hardware if, *e.g.*, it employs a complicated pipeline managed by sophisticated speculative algorithms. Technological advances, on the other hand, allow for immense processing power without resorting to such features, freeing chip space to be used for application-oriented on-chip resources.

In the sequel, the undesirable properties found in conventional architectures will be pointed out. Further, a non-orthodox processor architecture will be outlined, which is able to cope with non-deterministic occurrences of events. Finally, an experimental implementation of it will be sketched.

3.1 Undesirable Properties of Conventional Hardware Architectures and Implementations

In hardware architecture design, an unfortunate situation can be observed: although technological advances offer immense possibilities for design and im-plementation of processor architectures suitable for embedded real-time appli-cations, the processors produced are mainly designed with universal comput-ing in mind. Except for simple microcontrollers with low performance and only basic features, modern processors exhibit serious drawbacks when employed in embedded real-time applications. The main reason for the poor suitabil-ity of current microprocessors for real-time processing is the mismatch of the global objectives in the design of universal processors and those dedicated to real-time applications.

Besides technological advances, architectural means have always been em-ployed to increase the performance of microprocessors. Examples are pipelin-ing, caching, or virtual addressing. Virtually all of these measures try to op-timise the behaviour of processors and computers in the most common sit-uations of computational processing. Embedded computer systems, however, operate in real-time mode. Most often, its requirements are demanding, and it is of utmost importance that tasks are completed within the requested time frames, implying consideration of worst-case behaviour in contrast to average-case behaviour. This is the main mismatch in the design requirements holding for generic devices on one hand and created by real-time environments on the other. No high performance processor enables the exact calculation of a program's execution time even at the machine code level, as is possible for the 8-bit microprocessors developed the 1970s.

For that reason, we are forced to attempt other approaches to obtain worst-case execution times in absolute units. Some of them will be presented in Section 4.2. In the immediately following sections, certain measures to improve average performance of computer systems will be dealt with. It will be shown how they influence temporal behaviour and predictability of task execution.

3.1.1 Processor Architectures

The increase in microprocessor performance was, to a great extent, influenced by novel ideas employed in architectural design. These ideas were the RISC philosophy, parallel processing (pipelining), and caching. Measures related to the RISC idea also appear to be very useful for architectures employed in hard real-time systems, because of the simplicity and verifiability rendered. The absence of a microprogrammed control unit promises simpler, more transparent instructions and corresponding execution. However, in conjunction with other measures introduced at the same time, the RISC philosophy did not do much for worst-case run-time estimation.

Two further performance enhancing features are a high degree of internal parallelism in the execution of instructions, including the pipelining concept, and caching. Should new processors explicitly dedicated to embedded real-time applications be designed, these features ought to be used only with extreme precaution — if at all.

Independent Parallel Operation of Internal Components and Pipelining

In order to exploit fully inherent parallelism of operations, the internal units of processors are becoming more and more autonomous, resulting in their highly asynchronous and independent operation. There is no more microprogrammed control unit that would sequentially invoke certain units with known response times, as was the case for 8-bit and early 16-bit microprocessor families. It was common to have execution times of each instruction, or at least minimum and maximum figures, listed in manuals. This is no longer possible, since the execution times mainly depend on the sequence of instructions within a program. Moreover, to utilise the parallelism of internal processing units, the programs are analysed and instructions are executed out of order, utilising certain heuristics and speculations.

By this approach, the average performance of processors is increased. However, the complex task remains to analyse corresponding sequences of machine code instructions in order to determine their execution times, especially bearing in mind that the microprocessor vendors do not supply explicit and in-depth information about instruction execution. Also, for the time being, verification of such complex processors is impossible. Thus, it is questionable if they can be permitted to be employed in extremely critical applications.

The most common way to improve a processor's throughput by exploiting instruction-level parallelism is pipelining. Apart from the fact that by using a pipeline the execution times largely depend on the sequence of instructions, there are other effects that jeopardise the predictability of program elaboration. In particular, the well-known pipeline breaking caused by run-time decisions (conditional branching) based on sophisticated prediction methods is difficult to be assessed in advance.

Various techniques are used to cope with the pipeline breaking problem due to the dependence of instruction-flow control on the results generated by previous instructions. One of the oldest and most common techniques is the *delayed branch*, *i.e.*, the instruction immediately following a conditional branch instruction is always executed. It is a responsibility of the compiler to place an instruction there, which needs to be executed in any case, or, if impossible, to insert a no-operation instruction. Thus, if branches appear frequently in a program, which is the most realistic presumption, the principle of delayed-branch pipelining provides for predictable operation, but little gain in performance, since a lot of idle instructions needs to be inserted.

Another measure to improve the performance of pipelines is *branch prediction:* guessing ahead of time whether a branch instruction is going to be successful and whether the jump will be performed or not. If the guess is correct, then the pipeline will remain intact and fully utilised. If the guess turns out to be wrong, then some steps must be discarded and the pipeline will be less efficient. Statistically, it is possible to predict correctly with a probability of 90%.

There are two main approaches to branch prediction. In *static branch prediction*, one of the alternatives is selected always to be tried first. Which alternative that will be can be determined off-line by the compiler or directly using some heuristics (*e.g.*, branching backwards is more frequent in the case of loops, thus more probable and always guessed first; as branching forward, however, is used to leave loops, it is more probable that such jump will not happen and, thus, the subsequent instruction is tried). This approach yields shorter delay times if the guess is wrong. For *dynamic branch prediction*, a monitor observes and compares the behaviour of branch instructions in the past. The more frequently taken alternatives will be tried first in the future. This solution requires additional hardware resources, and may produce longer stalls in the pipeline if the alternative taken first is wrong. More sophisticated architectures utilise higher levels of parallelism. In the case of a branch instruction, the pipeline splits, and both the target and the next subsequent instructions are elaborated. The pipeline of the instruction not taken is simply discarded.

Based on detailed information about instruction execution and pipeline prediction mechanisms which are, unfortunately, often proprietary, execution times could be predicted by complex machine code analysers or other approaches, *e.g.*, simulation of system behaviour. Owing to its complexity, a processor with pipelined operation is also very difficult to verify. Thus, it is recommendable either to prevent pipeline breaking, or to renounce fully pipelining in the eventual design of processors dedicated to embedded real-time applications.

Another problem, which is particularly relevant for real-time systems, arises, when a task is pre-empted. Since one program's execution is then discontinued and another one started, the pipeline is discarded, and the new target address is fetched into the pipeline(s).

Caching

Fetching data from off-processor sources represents a traditional bottleneck, especially in the von Neumann architecture. To copy with it, cache memories were introduced: a copy of the data read or written over the system bus is saved in the cache in order to have the data available when needed again in the near future.

Caches can be used for both program code and data. If programs consist of relatively short loops, caches yield rather good results. In the case of long loops or longer straightforward sections of programs, however, cache data need to be partly overwritten. That means that code will have to be re-loaded from program memory. This similarly holds for data caches that will be overwritten if there are a lot of data involved. Since the difference between the times to access data in memory and in caches is significant, program execution times thus depend greatly on whether instructions or data can be found in a cache or not. This effect cannot be neglected in timing analysis of software running on sophisticated target architectures. Knowing the behaviour of the compiler, and/or the generated object code, it should be possible by a sophisticated analyser to estimate whether data will be found in a cache or not.

Cache hits or misses depend on how programs and data are mapped to memory. Dependencies can relate to a whole process or even several process executions. This makes simulation or tracing particularly difficult, as users cannot anticipate the effects of address mapping in compilers and linkers and, therefore, cannot provide the necessary simulation patterns to test critical cases. Even the most sophisticated analyses would fail in the case of multiprogramming, where caches are filled with new contents on every context-switch. Cache misses can be reduced if future data access patterns are analysed. A block that is found to be used again in the future should not be written back by the cache replacement strategy. This approach, however, requires complex data and program flow analyses of the applications.

Thus, to predict worst-case execution times, it is safest to take into account the time required to fetch instructions from memory *via* the system bus. This means that the effect of caches is renounced. In the design of new architectures for embedded real-time applications it is, thus, not reasonable to employ the cache mechanism at all.

Registers

Traditionally, processors perform operations on data in registers. Early processors only had a few registers or accumulators, some of which were dedicated to specific operations, resulting in a lot of data moving between the registers, and between them and memory. Later, technological advances provided enough space for register files, *i.e.*, larger groups of fast accessible common-purpose registers. These could be utilised even by high-level language programming through specific data types declarations. The use of register files is convenient for real-time applications, since the times for data access are deterministic. However, if an application requires more data space than available in a register file, the latter must (at least partly) be used for temporary data storage for processor functions, *e.g.*, arithmetic operations.

Further, when re-entrancy is required, variables cannot be statically mapped onto registers, but have to be kept dynamically on a stack in main memory. Also, to achieve compatibility with previous processor versions, often compilers may renounce the use of available register sets for direct variable storage (unless explicitly required), and keep their variables in memory. Thus, a lot of unproductive moving of data between registers and storage must be performed. Although behavioural predictability (apart of the caching effects discussed above) is not necessarily endangered, differents kinds of on-chip temporary data storage should be used to present the mentioned inconveniences.

In hardware technology, there is the unfortunate situation that processor speed increases by 60% per year, while data latency improves only by 7%. Thus, dozens of processor cycles are needed to supply operation arguments from memory. Expensive off-chip and on-chip static memory caches, that are supposed to deal with this problem, have already been criticised above. On the other hand, technological advances allow for large memories to be implemented directly on processor chips. In the design of novel processors for embedded real-time application, this possibility should be explored. It is in line with the idea of an "Intelligent RAM" (IRAM) [90], *i.e.*, a processor surrounded by vast areas of on-chip DRAM where both data and program memories could be accommodated.

When data are available on-chip in significant amounts, there are no more delays caused by their transfer on external buses. Thus, direct access from the processing units to the data is enabled, possibly even supported by multiple data paths for parallel access to two or more operands. As another consequence, caches with their above-mentioned problems become superfluous. Further, the number of processor pins could be drastically reduced, since only small I/O spaces need to be addressed by a few address and data lines.

3.1.2 System Architectures

Direct Memory Access

Since general processors are ineffective in transferring blocks of data word by word under program control, direct memory access (DMA) techniques were designed as another measure to improve the performance of computer systems. With respect to the delays caused by DMA transfers, there are two general modes of DMA operation, *viz.*, *cycle stealing* and *burst mode*. A DMA controller operating in the cycle stealing mode is literally stealing bus cycles from the processor. Thus, during a DMA cycle in cycle stealing mode, the processor cannot access the bus and has to wait. An advantage of this mode is that the processor has control over the bus (at least) every second cycle, and can react to exceptional situations by stopping the DMA transfer if necessary. A drawback is that bus arbitration takes its time, which means that this overhead needs to be taken into account. In the early days of microprocessing, cycle stealing directly speeded up operation, since data processing was the bottleneck due to low processor performance and low clock frequencies in comparison with the bus bandwidth. In the meantime, however, the situation has changed: we experience immense increases in processing power, while the data transfer over buses is more or less unchanged and essentially limited by the laws of physics.

In burst mode, the processor is stopped completely until a DMA transfer is completed. Although the processor has no control over its system during such a delay, this mode appears to be more appropriate when predictability is the main goal. If block length and data transfer initiation instant are known at compile time, the delay can be calculated and considered in estimating a program execution time. The same prediction would be much more complicated for the cycle stealing mode. Furthermore, block transfer is also faster, because bus arbitration needs to be carried out only once. However, precautions have to be taken to enable a processor, whose operation was suspended for DMA operation, to react in the case of catastrophic events.

Virtual Addressing

To cope with a limited amount of memory available in computer systems, virtual addressing techniques have been developed. If there is not enough room in main storage, blocks of the memory are kept on disk and only the ones of actually executing programs are resident in physical memory. Once a part is not needed any more, it is swept back to disk and replaced with those of the newly started programs. Using this technique, data access times are very much dependent on whether requested items are already in memory or pages have to be loaded first. To determine this beforehand, a complex program analyser would be necessary. Even then, restrictions would have to be imposed. For instance, register-indirect addressing modes should only access data in the same page, which is difficult to assure when using a high-level programming language. It has to be noted, however, that due to the advances in solid-state technology, in most cases it is easy to provide enough memory for virtually any needs of embedded applications. Thus, it can be concluded that virtual addressing is obsolete and should be renounced for hard real-time applications.

Data Transfer Protocols

Data transfer also deserves some attention. In this paragraph microcomputer buses are considered. Data transfer in local area networks will be discussed later in Section 3.5.

As already noted, the lowest-level physical speed on computer buses is limited by physics. There are some hundred data lines running close to each other, with signal frequencies of several hundred Megahertz and all their negative effects such as capacitive losses, crosstalk, resonating, reflected and standing (stationary) waves.

At higher levels, data transfer is controlled by transfer protocols, which provide for communication and synchronisation between sender and receiver. There are two classes of protocols, synchronous and asynchronous ones. By synchronous protocols, all units operate with the same clock, which is also transmitted or synchronised over the bus. In the early days of microprocessors, this was the only feasible way due to its simplicity. Later, it was noticed that due to very diverse operating speeds of different units on a bus, it was not optimal and that more flexible ways of synchronisation were needed. This was provided by asynchronous data transfer based on handshaking between a master and slaves in communication. No common clock is needed any more, and each unit operates at its own speed. Also, adaptation to new units in a system is transparent. A considerable amount of time is, however, lost to the handshaking protocol.

Later, due to technological advances, the situation changed in favour of the synchronous protocols again: the speeds of the units became comparable.For this reason, the more effective synchronous protocols found their way into RISC processors.

By definition, synchronous data transfer protocols ensure predictable data transfer times, whereas the behaviour of the asynchronous ones is very difficult to control, especially in shared-bus systems. Thus, synchronous communication provides an easy way to guarantee realistic data transfer times. Similar conclusions hold for local area network protocols, which should be based on synchronous time-driven protocols.

3.2 Top-layer Architecture: An Asymmetrical Multiprocessor System

An important issue to be pointed out when discussing temporal behaviour of embedded real-time systems is the use of signals from the environment and other events that interrupt normal program flow as the method of invoking processes that need to respond to signals from the process environment. This approach is widely used and appreciated, since, unlike polling (interrogation of environment variables) and busy waiting, it offers best processor utilisation and is problem-oriented.

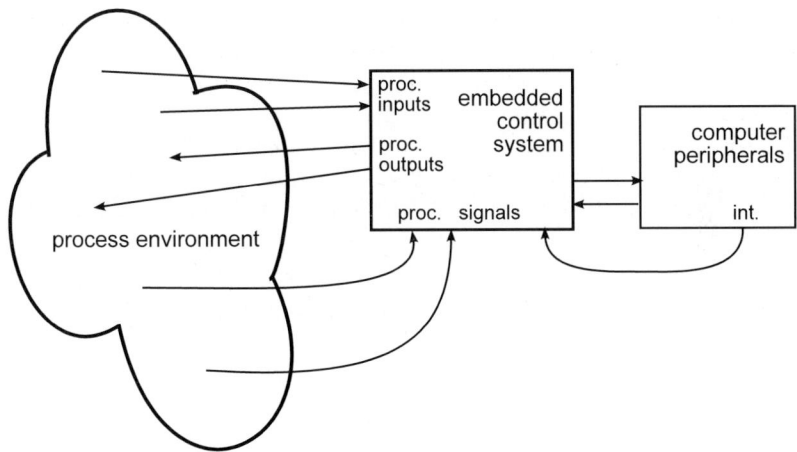

Fig. 3.1. Interrupt concept

In Figure 3.1, the control application running in the control system is interfering with the process environment through process inputs and outputs. The data exchange is controlled by and, thus, synchronised with the program: its duration and delays are considered in the estimation of the worst-case execution time. On the other hand, events reported from the environment to the control system by signals (interrupts) occur sporadically, and are not synchronised with program execution. Usually, immediate attention of the control system is required. This has been common practice in programming microprocessor-based control applications. There are at least two problems originating from this practice:

1. Interrupt servicing is an event-driven approach. Unlike the state-driven one, there is always a danger of missing an event, if, *e.g.*, the interrupts are masked for some reason. If such an event changes the state of the system, that could represent a serious error. If, on the other hand, in the case of state observation a change is missed, this will probably be noticed during the next poll. For detailed elaboration *cf.* [71].

2. More severe, however, is that interrupt or exception servicing necessarily delays the execution of the running task. It is beyond the control of the scheduler or the task itself, so it cannot be taken into account during schedulability analysis. If the feasibility of a schedule has been proven, it may not hold any more after a non-anticipated and non-predictable delay due to a sporadic interrupt.

Thus, it is necessary to introduce an architectural measure in order to deal with signals from the process environment, which request service by the control system.

In this section a possible solution will be shown, based on a distributed asymmetrical multiprocessor architecture. The original idea was created in the late 1980s; similar approaches can be found in a number of contributions [47, 108, 79, 103, 22]. It is not the purpose of this section to propose an ultimate solution to the problem of designing a consistent real-time processor architecture, but to launch some ideas for further elaboration.

3.2.1 Concept

As is common in engineering, there are always many possible system designs fulfilling a given set of demands — provided the problem is solvable with available technology. To derive such an appropriate architecture, we start off with a basic reasoning.

First, homogenous symmetric multiprocessing, where a task may be dispatched to run on any of the processors, is not realistic in process control. At least in the case of front-end processors with process interfaces usually physically wired to sensors and actuators establishing the contact to the environment, it is natural to implement either single-processor systems or dedicated multiprocessors acting as separate units. Further, tasks may require specific processor features like signal processing or floating-point arithmetic co-processing which are not always available at all processing nodes. Thus, asymmetrical multiprocessor architectures with dedicated processors for specialised task processing are more reasonable.

The above reasoning advocates against dynamic mapping of tasks onto more uniform processors in order to utilise better them. Scheduling of such systems would be possible (e.g., using the least laxity first strategy) but schedulability analysis would be much more complex. As shown in Figures 2.10 and 2.11, the context-switching overhead is much higher than that caused by the earliest deadline first approach; the latter, however, does not support dynamic multiprocessor scheduling in a feasible way. If, however, certain tasks do not require any specific processing or other facilities, their ability to be statically re-mapped onto different processors may greatly enhance system robustness: if a processing node fails, the system can be re-configured by mapping the tasks on the remaining nodes. In this event, re-scheduling is necessary. The problem of doing that seamlessly remains an open issue.

Further, let us consider an analogy from another field, where systems coping with real-time conditions have long been developed and used. The example system consists of a manager and his secretary. The duties of the secretary are the reception of mail and telephone calls, the elimination of unimportant chores, and the minimisation of interruptions to the manager's work by visitors and callers. Furthermore, she schedules the manager's work by arranging the files in the sequence in which they are to be deatl with, and through the administration of his meeting appointments. Thus, the manager's work becomes less hectic — *i.e.*, the work's "real-time conditions" are eased — and more productive because he can perform his tasks with less frequent interruptions in a more sequential and organised manner.

By taking patterns from this and related models, we now define the overall structure of a computer designed to meet the requirements of real-time operation. In the classical computer architecture, the operating system is running on the same processor(s) as the application software. In response to any occurring event, the context is switched, system services are performed, and schedules determined. Although it is very likely that the recent event was of lower importance and the same process will be resumed, a lot of performance is wasted by superfluous overheads. This suggests employing a "secretary", *i.e.*, a parallel processor, to carry out operating system services. Such an asymmetrical architecture turns out to be advantageous since, by dedicating a multi-layer processor to the real-time operating system kernel, the user task processors are relieved from any administrative overhead.

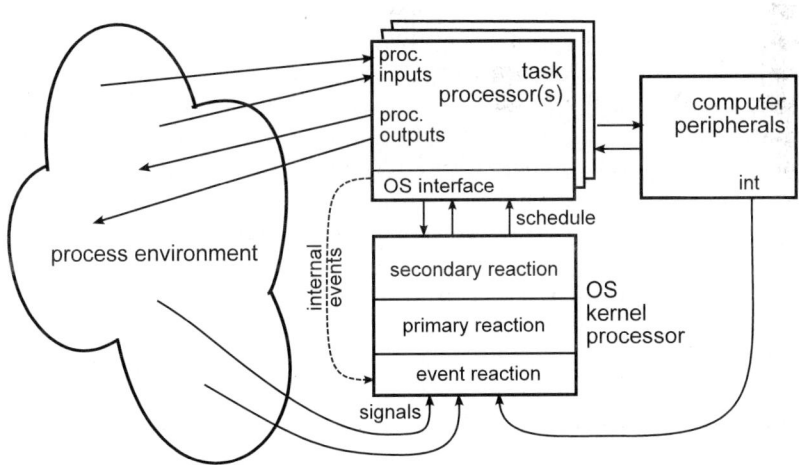

Fig. 3.2. Conceptual diagram of the asymmetrical multiprocessor architecture

The asymmetrical concept is displayed in Figure 3.2. It has been presented and elaborated in detail earlier [47]. The computer system to be employed in hard real-time applications is organised hierarchically. The basic system consists of two dissimilar processors, the task processor and the operating system kernel processor, which are fully separated from each other.

The *task processor* is usually a classical von Neumann processor executing control-application tasks. It acquires sensor information from the controlled environment, and controls it by actuation output generated. The task processor also executes a small part of operating system tasks. There is a dispatcher responding to re-scheduling requests, and certain functions that interface to user tasks. Specifically, the operating system tasks on this processor are mainly outer supervisor-shell services, such as data exchange with peripherals and file management (if required), provided in the form of independent tasks or subroutines called by the user tasks. An example of such a processor and its design guidelines were elaborated in [16, 17]. Being the creative part, the task processor corresponds with the manager in the above analogy.

On the second processor, the dedicated *operating system kernel processor*, which is clearly and physically separated from the outer-layer tasks, operating system kernel routines are running. This co-processor houses the system functions event, time and task management, communication and synchronisation. Although important and actually controlling the operation of the task processor, these functions are routine ones and would impose an unnecessary burden to the latter; thus, the kernel processor corresponds to the manager's secretary.

If necessary, this concept can easily be extended to multiple task processors, each one executing its own task set, but being controlled by a single operating system kernel co-processor.

The operation of the whole system is shown in Figure 3.2: the external process and the peripherals are controlled by the task processor *via* I/O data communication. To prevent delaying task execution by immediate event handling, they are fed into the lowest, the event reaction layer of the operating system kernel processor. This way, the task processor is only interrupted by the kernel processor if an event is relevant and a context-switch is really necessary. As events, all sources requiring attention by execution of service routines are considered. These are external events, which originate in the environment and are transmitted as signals, temporal events, or events generated by the task processor, like synchronisation or explicit software events.

The migration of event handling to a parallel processor is the most important feature of this approach allowing for temporal predictability of application task programs and their run-time schedulability analysis. While the task processor(s) execute task programs, event administration, scheduling, and schedulability analyses are carried out in parallel without producing any delays. All this is done upon the run-times, deadlines and other information on tasks kept in Task Control Blocks (TCB). Their system parts reside in the operating system kernel processor, while the contexts, containing the initial states of application program and the states of pre-empted tasks, are stored in the task processor(s).

Communication Between Task Processors and Operating System Kernel Processor

Communication and connection links, over which the operating system kernel processor and the task processors communicate with each other, can be implemented in different ways. When choosing a suitable implementation, it is important that the traffic on the link from a task processor to the kernel processor is not influenced by links connecting other task processors and is, thus, not non-deterministically delayed in a case of demands occurring with high frequency.

For instance, in an early implementation [18], the authors used a transputer as kernel processor, and its serial point-to-point links to communicate with the task processors. This way, independence of links was inherently assured by separate link interfaces on the transputer. However, the number of task processors was limited to those of links, *i.e.*, four. Communication was implemented in form of two serial bus lines, the *data* and the *sync* line. Data on the data line were confirmed by the transmitter with a strobe on the sync line. Each partner could transmit a packet of data if the bus was idle, which was represented by high state of the sync line. The partner wanting to prevent communication forced the sync line low, thus disabling the initiation of transmission by the other partner.

Through the connection links, the task processors request certain services and provide different information on one hand, and the kernel processor requests context-switches and delivers required information on the other. In the communication protocol, the task processors have the ability to postpone communication with the kernel processor for a short time while executing tasks are inside of non-pre-emptable critical regions.

When the kernel processor requests tasking operations such as task initiation or pre-emption, they are performed by a small operating system kernel counterpart, a task dispatcher, which is running on the task processor(s), and which also administrates the application part of the tasks' states kept (mainly their current contexts) in the application parts of the task control blocks.

3.2.2 Operating System Kernel Processor

The functions of the kernel processor are, in short: (1) to receive and recognise all kinds of events, and to react to them by changing task states, and (2) to provide other operating system support to task processors. In Figure 3.3, the functions and communications between the layers of the operating system kernel and task processors are summarised.

The operating system kernel processor provides support for the following functions:

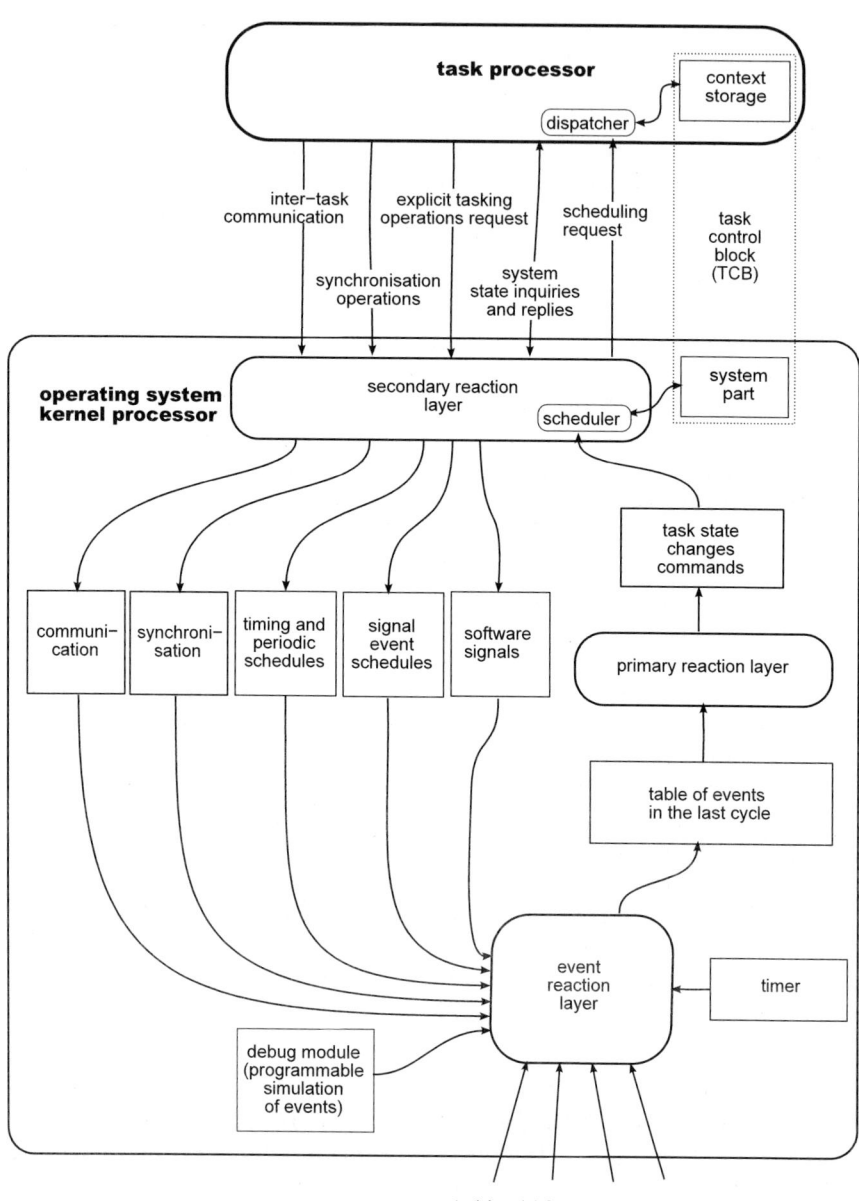

Fig. 3.3. Functional units of operating system kernel processor

Task scheduling. The kernel processor's essential function is scheduling of tasks to be run on the task processors. For this, the paradigm earliest-deadline-first is employed. Upon any event in the environment and in the system, the task execution schedules for all task processors are generated. At the same time, schedulability analyses are performed to verify whether the schedules are feasible, considering the remaining execution times of all previously scheduled tasks and newly arrived ones, together with their resource requirements and other limitations. If feasibility cannot be acknowledged, fault-tolerance measures must resolve the situation, possibly by re-configurating the tasks in the sense of graceful degradation, or by other suitable means.

System state access. Tasks may request information about the system state (*e.g.*, time or state of synchronisers), which is made available through monadic operations in form of system calls.

Explicit task control commands. A task, running on a task processor, may need to perform certain tasking operation, like activation, termination or resumption of another task, scheduled for certain events or combinations thereof. To allow for this, tasking commands can be submitted to the kernel processor. Moreover, a task can also trigger certain events synchronously to its execution. Tasking operations may be scheduled as response to this kind of software signals.

Synchronisation. A further function of the kernel processor is task synchronisation, for which it provides the adequate operating system kernel routines with synchronisers and waiting queues. Having exclusive access to the synchronisation primitives, it can perform atomic actions upon them, thus preventing race conditions. By managing the waiting tasks' queues, it can supervise the corresponding operations upon tasks like suspension and resumption. It can also periodically check the temporal bounds of suspended tasks, and perform the actions related to time-out conditions, *cf.* Section 2.3.6.

Inter-task communication. While the communication itself is implementation-dependent and will be elaborated later, there may be events triggered by communication, like, *e.g.*, arrival of messages, that need to be supervised by the operating system kernel.

The kernel processor consists of the three hierarchical layers detailed below. Through their asynchronism and their loose coupling by data tables, this functional decomposition also suggests their parallel operation and the technical implementation of the operating system kernel processor as, *e.g.*, application specific integrated circuits.

Event Reaction Layer

On this layer, events of all kinds are recognised and forwarded in a uniform way to the primary reaction layer. The events are either received from other sources, or generated based on observation of certain system data structures.

Apart from that, the layer also performs certain low-level functions like timing and start-up initialisation:

Handling of time. A hardware timer/counter-based high-resolution clock is maintained, giving pace to the entire system. Considering the fact that in the case of malfunction this can present a serious potential central point of failure, the implementation must be adequate; here it is referred to the details of a proposed implementation by low-level system re-configuration (see description on page 95).

External events (signals). When events from the environment are signalled, they are latched together with the times of their occurrence (*"time-stamps"*), and made available to the primary reaction layer. The timestamps provide information necessary for the scheduler to establish their deadlines: the response time of a task must be counted from the instant of occurrence of the initiating event. Another important related feature is monitoring of event overrides: if an event appears before its previous instance has been handled by the primary reaction layer (*i.e.*, transformed into a task state change command), this violation is reported to the primary reaction layer in the table of events. Where applicable, the same holds for all other kinds of events. Timestamps and override information is kept for all events and not only external ones.

Time events. If there are tasks that have been delayed or put in periodic execution, their schedules are kept in tables accessible by the event reaction layer; their waiting times are decreased and, upon expiration, events are generated.

Synchronisation events. On changes of synchroniser states, the corresponding waiting queues are scanned and events triggered, *e.g.*, when a condition for a task to resume to the ready state is fulfilled.

Communication events. Similarly, events may be triggered on certain state changes in communication, *e.g.*, when a message arrives.

Programmed events. There is the possibility that a task may trigger a software event (in a form of a "software interrupt"), upon which an operation may be scheduled.

Programmable event generator. For software simulation purposes, a separate programmable interrupt generator is provided.

Application-specific events. Sources of such events may be introduced if necessary.

To summarise, in this layer basic administration is performed, and all kinds of events are brought into a common format. Together with their arrival times, they are forwarded to the primary reaction layer in a table of events that occurred after the last servicing. This table also keeps track of possible overrides of non-serviced events, *i.e.*, if an event happens before its last occurrence is served.

Primary Reaction Layer

The purpose of this layer is to map events to operation requests. The according commands to change task states are issued and forwarded to the secondary reaction layer. The layer functions in a way similar to a programmable logic controller, *i.e.*, as a cyclic executive performing only one task. Within every cycle the possible sources (either internally generated time events, changes of synchronisers, or external events) represented by the event reaction layer in the form of table elements are polled.

In the applications, tasks and tasking operations are connected to certain events by language constructs. It is the function of the primary reaction layer to map and to trigger the appropriate tasking operation by putting task into certain states. For example, a task activation may be triggered by a hardware signal from the process environment, or a periodically executing task may be terminated after a specified number of repetitions.

In summary, the primary reaction layer converts the event-driven asynchronous operation of the event reaction layer into periodic state observation.

Secondary Reaction Layer

This layer performs the higher levels of the operating system kernel. It provides an interface by communication routines that receive and transmit messages to and from the task processor(s). By those, tasks running on task processors are requesting services from the operating service kernel routines. These are related to tasking, synchronisation, communication, and other commands; the kernel processor responds to these requests by performing the services or providing replies. In addition, as its most important function, the secondary reaction layer performs scheduling algorithms in reaction to events handled by the event reaction layer and interpreted by the primary reaction layer. By specific messages, requests for pre-emptions and dispatching are submitted to the task processors.

In safety-related environments, establishing and checking initial configurations is necessary. This is another task of the operating system kernel processor. After power-on, the initial auto-diagnostics protocol is started. All units perform self-check procedures and gather, in a recursive way, information on the subordinate units they are responsible for. So, *e.g.*, task processors check for availability and correctness of the peripheral interfaces needed, and the kernel processor verifies functioning of the primary and event reaction layers. Self-checking can be carried out on the level of hardware (*e.g.*, voltage level check, power consumption, boundary scan features), or by software (running diagnostic procedures like memory tests or loop-back mode testing of communication systems.)

After its self-check, the secondary reaction layer establishes communication with the task processors, and by polling verifies their presence and flawlessness. The redundant components are detected and, based on this information, the initial configuration is selected from a predefined set. If certain components are missing from the beginning, the objectives may be modified and the application runs in a degraded mode. During this phase all necessary initialisation actions, like synchronisation of the clocks, are performed Finally, the system data structures needed by the event and primary reaction layers are created and initialised, and the application is started.

3.2.3 Task Processor

In the task processor(s), the application processes are executed without overhead due to operating system functions. They interact with the controlled process by their inputs and outputs *via* peripheral interfaces, synchronously to program execution. No process signals or other asynchronous events are handled directly by the task processor. Thus, its operation is deterministic and provides for predictability of task execution. The latter is initiated by the kernel processor, and is only pre-empted if necessary, to assure meeting the deadlines of processes, according to the scheduling policy.

The processors used in embedded systems for task program execution are usually general-purpose microprocessors, microcontrollers or digital signal processors. Except for the most simple ones, they do not exhibit temporal predictability, which is the pre-condition for microscopic (or layer-by-layer) predictability. The reasons have already been dealt with in detail in Section 3.1. In summary, they are due to employing processors developed for non-real-time applications. Their design and the optimisation guidelines used are based on statistical average performance and not on their worst-case behaviour. Actually, there exist virtually no high-performance processors that would provide deterministic or bounded worst-case temporal behaviour. This unfortunate situation led us to experiment with a possible hypothetical architecture that should be suitable for embedded real-time applications. Without going too deeply into computer architecture, or even microelectronics, we have set the following goals:

- To explore which features should guide the design of an architecture characterised by deterministic instruction processing,
- To exploit the technological advances, most notably the immense number of transistors on a chip, and
- To support parallel execution of control and data processing instructions, exploiting the natural parallelism.

The resulting hypothetical processor architecture was elaborated in detail in [16, 17]; it is outlined below.

As suggested in Figure 3.4, functionally, the task processor is divided into the *task control* and the *data processing* units, residing on the same chip. Program and data memories are separated in the form of the Harvard architecture. The control unit accesses the program memory, only, and the data processing unit only the data memory. Instructions of different kinds are executed at the most appropriate places. Flow-control instructions are processed close to program memory, where the program counter is also located. To the data processing unit, only instructions dealing with data are sent in a straightforward instruction stream, without any program addresses, thus reducing the number of data transfers. Process data are exclusively handled in the data processing unit from where external sources are accessed.

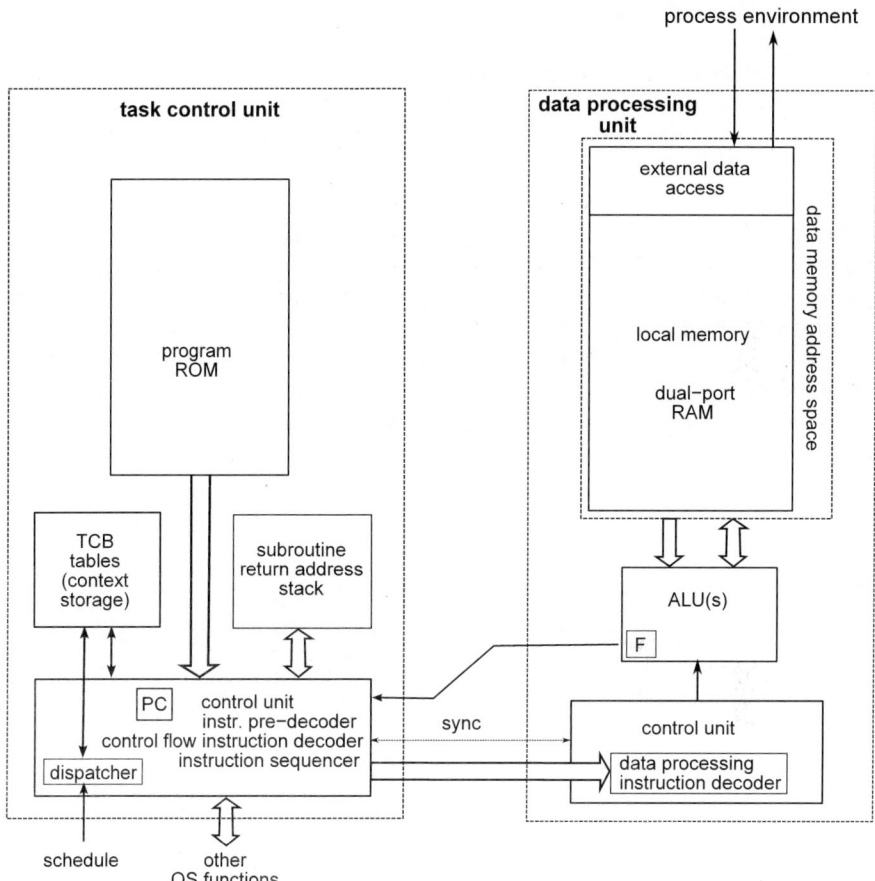

Fig. 3.4. Structure of the task processor

Task Control Unit

This unit's function is to execute task programs, which are kept in ROM for safety reasons to make self-modification of code impossible. From this program memory instructions are fetched and pre-decoded to distinguish between flow-control and data processing instructions.

Only the *program-flow-control instructions*, which have an impact on the program counter, only, are executed here. To allow for conditional branches, a status flag (F) is accessible from the ALU of the data processing unit.

For the administration of subroutine return addresses a stack is provided. Because of the pre-emptivity of tasks, strictly nested operation is not assured. Thus, each task needs a separate stack; its stack pointer is kept in its TCB. The amount of stack memory, however, is limited, because recursion is not allowed, and asynchronous events are not handled here. Thus, the depth of procedure nesting and the stack memory necessary can easily be determined in advance by the compiler; also, by the latter, initial stack pointers are calculated and stored into the TCBs.

The *data processing instructions* are forwarded to the data processing unit. While they are being executed there, the synchronisation line indicates the busy state, preventing the task control unit to send further data processing instructions. However, flow-control instructions can be executed in parallel. This distribution of instruction execution minimises unproductive data transport. The program counter (PC) is exclusively administered in the task control unit, also eliminating the need for program address transmissions. This way, the inherent parallelism is utilised, yielding similar effects as pipelining, however in fully deterministic way.

Data Processing Unit

The data processing instructions forwarded by the task control unit are decoded and executed here. For that purpose, the necessary arithmetic and logic units are provided, addressing data in the unit's data memory address space. The latter consists of two parts, local memory and external data access space. Instead of registers, the *local memory* appropriately supports the variable concept of high-level programming languages. Since advances in technology allow for a large number of transistors per dye, part of them should be used to implement local storage. It keeps all local variables and constants as occurring in high-level language programs. Each task has its own statically defined portion of the local memory.

To prevent problems with dynamic allocation of data storage, recursion should not be allowed; in the very rare cases that it occurs in process control applications it can easily be substituted by iterations. Thus, variables can be statically mapped onto locations in local memory.

Constants can be handled in several ways. Standard constants like 0 and 1 are statically implemented, and available to all programs at specific dedicated addresses, as is common in RISC microprocessors. Other constants can be stated in the processing instructions (equivalent to the addressing mode immediate): this part of an instruction is interpreted as a literal and used as operand. As another possibility, the compiler could substitute numeric constants by variables in local memory. Adequate values are loaded into those variables at initialisation time.

To enable dyadic operations, special design of the local storage (synchronised dual-port access) allows for fully synchronous reading of two different locations. In writing a single location is always accessed and at different instants than reading thus no conflicts are possible.

The memory locations of the *external data access space* are accessible *via* a special portion of the data memory address space. By external data, global common memory and memory-mapped peripheral interfaces are meant. There are various different possibilities for implementation; the two most characteristic ones are indirect addressing in global memory, and distributed replicated shared memory.

Indirectly addressed global memory locations and peripheral device registers are automatically accessed *via* pointers in local memory. When the dedicated portion of the data memory address space is accessed, the contents of the local memory read from the addressed location represents *the address* of the operand in global data storage. These data are automatically fetched *via* external memory access and output to the ALU instead of the local memory contents. Except for the longer access time, the operation in the data processing unit is exactly the same as in direct addressing mode.

Distributed replicated shared memory is a portion of local memory, being a replica of a part of distributed global memory. Each task processor needing access to a certain global address space maintains its own copy of it, which is transparently updated by background mechanisms. On the other hand, if a processor changes the contents of its copy, the change is replicated in all copies. There is the problem of conflicts when more than one processor wants to change the same memory location; these conflicts must be solved by appropriate implementation measures.

To perform *arithmetic and Boolean operations*, an ALU is provided. The main requirement for its design is temporal determinism of operation. The basic arithmetic operations on integer and 32-bit floating-point numbers need to be provided. For floating-point implementation, the IEEE P754 and P854 standards were followed, including their exception handling.

In some applications, the transfer of run-time information between the task control and the data processing units could be useful (*e.g.*, to provide some system state information). For that purpose, optionally, a register for data-communication between the two units may be foreseen. Although useful, the implementation of such register implies, however, a data move instruction between the register and local memory, which has been avoided so far. Instead, the task control unit could provide this information as part of the data processing instruction (*i.e.*, in form of a literal constant in immediate addressing).

Instruction set

The architecture proposed above is simple and clear and, thus, promises easier verification than complex state-of-the-art processors. In order also to keep low the complexity of the control units of both task control and data processing units, a reduced instruction set appears to be the only reasonable choice. A further motivation is the possibility of the processor's implementation in ASIC technologies for which low complexity would be convenient.

Following the separation of instruction processing units, the instruction set is divided into flow-control and data processing instructions. The former are conditional and un-conditional branches, subroutine calls and returns, and the wait instruction, by which the processor's characteristic of time-awarenes is emphasised. Instruction sequencing is delayed for either a specified period of time, or until an event being awaited by the kernel processor occurs, whichever comes first. If the continuation signal from the kernel processor is received before the period expires, a branch is performed, else the program continues execution of the next subsequent instruction. This way, worst-case time-bound waiting for events can easily and effectively be implemented.

The three-operand data processing instructions perform basic arithmetic and Boolean operations on both integer and floating-point operands, depending on the arithmetic/logic units implemented. Addressing of two source and one destination operand by a single instruction renders registers superfluous and minimises unnecessary data transfers.

There are basically two data addressing modes, absolute and indirect. The absolute addressing mode refers to operands residing in local memory. By the indirect addressing mode, an operand is accessed in external storage *via* a pointer in local memory containing its physical address.

3.3 Implementation of Architectural Models

In this section, a brief history of ten years of implementing several variations of the above architecture is given. To represent the evolution of the ideas, three steps in the development are shown, together with their advantages and drawbacks that led to the respective next solution.

At this place, it should be noted that the model as described in the previous section is to be considered as a logical one only. For instance, the communication lines between task- and kernel processors (see Figures 3.3 and 3.4) represent logical communications of data, operating system routines calls, schedules, context-switching requests, *etc.* All these communications are implemented in the form of simple protocols on the data that are physically transmitted between the different units in the system. This transmission can proceed either on point-to-point connections or on a local bus, as will be shown below.

Most of the units of the kernel- and task processors are implemented in software running on microprocessors. In some cases, hardware implementations are feasible, *i.e.*, the prototypes on field programmable gate arrays and, possibly, in other ASIC technologies later. In this section, however, this will not be dealt with in detail.

3.3.1 Centralised Asymmetrical Multiprocessor Model

Let us start with an obvious implementation that, in a straightforward way, follows from the above architecture. It should be noted in advance that it may not represent the best solution. The implementation is shown in Figure 3.5. It uses several processors of possibly different types and performances, which implement the kernel- and task processors (and even intelligent peripheral interfaces, if necessary). Task processors are connected to the kernel processor by point-to-point links. This prevents possible collisions in task-kernel processor communication, which would jeopardise the predictability of the system's operation.

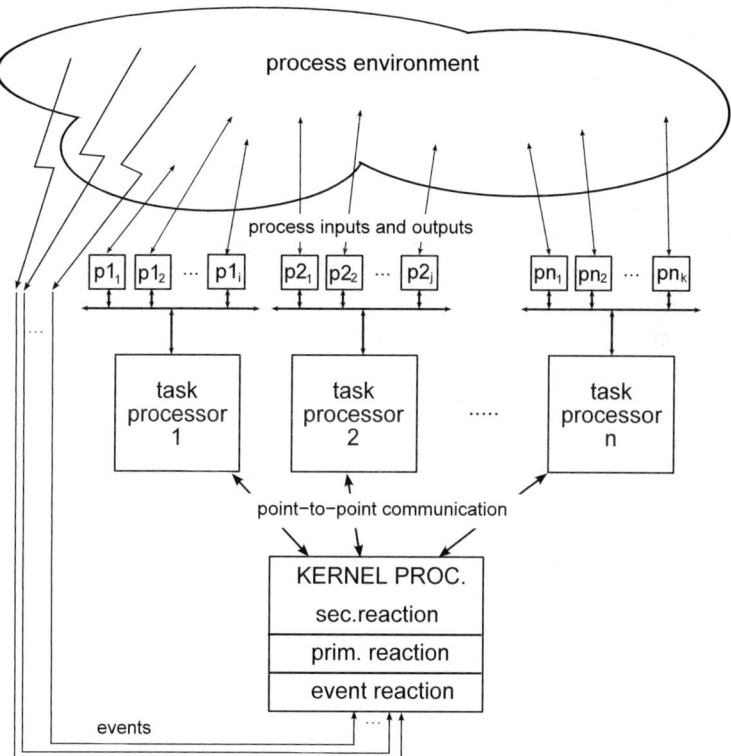

Fig. 3.5. Direct implementation of the asymmetrical multiprocessor model

Each task processor controls its set of intelligent peripheral interfaces in a master-slave mode: a peripheral interface is addressed by, and responds to, requests from the task processor. This way, there can be no collision on the local peripheral bus. The peripheral devices include a certain amount of intelligence; instead of being fully passive, they may perform some simple peripheral functions to relieve the processor from routine work, to provide some fault-tolerance features, and to be able to carry out some time-related operations.

Tasks are statically mapped onto the task processors. Since the peripheral interfaces are physically connected to them, this seems to be a reasonable choice. Further, the most suitable scheduling strategy, earliest-deadline-first, is not feasible when dynamically allocating tasks to different processors.

In the mid-1990s such an implementation was attempted in a prototype project [18]. It was built around a transputer as secondary (and simple primary) reaction layer, providing enough processing power for the rather complex operating system kernel with scheduling and schedulability analysis. For the event reaction layer (at that time called hardware layer), a Motorola microcontroller, enhanced by a Xilinx FPGA functioning as external event concentrator, was used. The transputer's four 20Mb/s serial links with very simple data transmission protocols were ideal for point-to-point connections to task processors. The latter were implemented with Motorola microcontrollers with CPU32 kernel. Each of them controlled a network of simple but, to a certain extent, intelligent peripheral interfaces, being capable of performing autonomous peripheral operations with temporal functions. They were connected using the simple I^2C communication (or Motorola's M-bus). The concept of distributed peripherals was adequate for process control interfaces. However, events still had to be transmitted on parallel connections from their origin in the process environment to the microcontroller performing the functions of the event reaction layer.

Since this was a serious drawback, in the next version all peripheral interfaces were residing on a common local bus which was connected to the secondary reaction layer of the kernel processor; see Figure 3.6. Task processors now communicated with their peripherals *via* operating system service calls, which presented somewhat more overheads. However, as there was no need to transmit very long messages from the task processors to the peripherals and *vice versa* due to the nature of process data, this was not a serious problem. Further, there was no danger of collisions on the local network, because it was controlled by the operating system service routines. Events in the environment were now observed by peripheral devices, providing for even more flexibility and more complex expressability of conditions that may trigger them. For each event, a bit in a special event message was transmitted to the kernel processor.

Communication was carried through cyclically using a dedicated nonstandard protocol which, in an interesting way, to a high extent resembled the time-triggered protocols that emerged a number of years later. The kernel

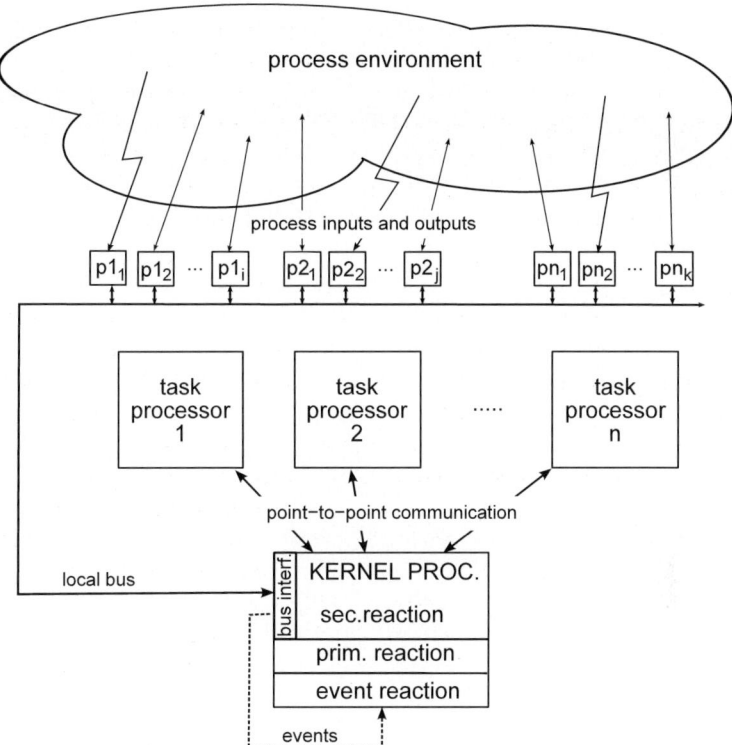

Fig. 3.6. Implementation of the asymmetrical multiprocessor model with peripherals on a local bus

processor periodically issued a reference message upon which the peripheral interfaces could synchronise their clocks. Then, the event messages were transmitted; if an event occured, the corresponding bit was set. An event message was a word accessible by all peripheral interfaces that could submit event service calls by connecting the data line to ground during its dedicated time slot. This message was received and decoded in the bus interface of the kernel processor, and events were forwarded to the event reaction layer. Event messages were followed by messages addressed to specific peripheral interfaces.

The implementation of the distributed peripheral interfaces had some other advantages in comparison with the centralised one. It is more flexible and allows for dynamic re-configuration of peripheral interfaces in case of failure. Also, it is possible to re-map tasks on task processors, since the peripherals are no longer physically attached to the latter. However, due to the selected scheduling policy, it was still not possible to do it dynamically during operation, but only upon system-reconfiguration in a case of failure.

Regarding fault tolerance, both implementations exhibited a serious drawback, *viz.*, a central point of failure: the kernel processor was implemented on a dedicated processor and could not be re-mapped in a case of failure. Further, the implementation was inflexible with regard to the number of task processors and their reconfigurability. Redundant resources other than peripheral interfaces could not be shared among the logical units to cope with failures. As a conclusion, although it was proven feasible, this implementation is not realistic. It conflicts with the common guidelines for implementing control applications, one of them being flexibility through fully distributed implementation.

3.3.2 Distributed Multiprocessor Model

The latter solution of the asymmetrical multiprocessor model where all peripheral interfaces are gathered on the same local network led to the general solution with the task processors also connected by the same local network see Figure 3.7. The obvious advantages of this solution are better flexibility, support for dynamic re-configuration, and fault tolerance:

- Task mapped to the task processors may be re-configured in case of a failure. Once this happens, the task set is again statically defined for the next period of time. True dynamic mapping of tasks on the task processors during operation, however, would be difficult because of the scheduling policy employed.
- There is a possibility for redundant units on the network in hot stand-by mode to be switched over in case of failure. These units simply access the input data from the network instead of the failed units and provide the results to the same users. The availability of redundant resources (task processors and peripheral interfaces) can be checked during the initialisation phase (see description of kernel processor).
- A central point of failure, as introduced by the centralised kernel processor, can be avoided by redundant resources available as stand-by units on the local bus.

By employing local-bus communication, a serious drawback has been introduced: problems with collisions on the network. Task processors operate asynchronously and could, thus, require services of the operating system kernel processor in the same instants. This, however, is a well-known problem of distributed embedded systems, operating in real time. It has been solved in different ways; the one we used — time-triggered communication — will be described in more detail later in this chapter.

3.4 Intelligent Peripheral Interfaces for Increased Dependability and Functionality

In this section, so far the operating system kernel and the task processors have been dealt with. Further components to be elaborated are peripheral interfaces. This subsection describes a suggestion to amend them with a certain

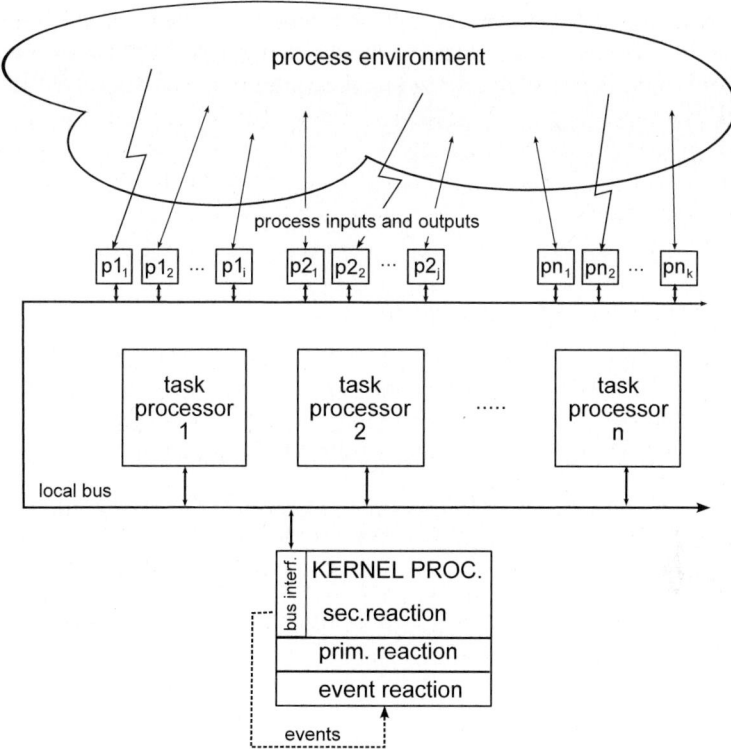

Fig. 3.7. Fully distributed implementation of the asymmetrical multiprocessor model

amount of processing capability. This way, a number of goals can be achieved:

Distribution of processing tasks. Data are processed locally in the parts where they are located or acquired. Thus, the main task processor's load is reduced, inherent parallelism in data processing is utilised.

Reduction of complexity of the application programs. Thus, they become less error-prone and more easily verifiable. Input data from the process environment are pre-elaborated, task processors receive higher-level information.

Reduction in data transfer. Not all raw data must be transmitted but only high-level information, which is normally less bulky.

Better tolerance to faults that originate in peripheral data acquisition and transfer in the distributed system architecture. The idea is to deal with the faults and resolve them, if possible at the lowest level in order to prevent the higher-level mechanisms to have to employ more costly measures.

De-coupling of peripheral devices from the task processor allows for asynchronism of peripheral operations; thus, they can be synchronised with the process environment.

To implement the intelligent peripheral interfaces, instead of passive memory-mapped peripheral interfaces as common in generic process control computers, we propose to employ independent units, which enhance system performance and robustness, and support consistent programming of temporal features. The process input and output interfaces are spatially distributed over the controlled process environment, close to the sources of data and actuators, respectively. To this end, simple microcontrollers were added to the peripheral interfaces. They can be pre-programmed to perform certain functions, thus relieving the general-purpose processor from detailed control of the peripheral interfaces, or from performing sequences of routine I/O operations. The required intelligence may also be implemented by application-specific integrated circuits. Such hardware-based automata are much easier to verify in a well-established way. Also, in certain cases, performance could be improved this way.

3.4.1 Higher-level Functions of the Intelligent Peripheral Interfaces

There are many ways to utilise the processing facilities of the peripheral interfaces. The most obvious is, instead of only acquiring sensor data or outputting actuator control data, to perform specific functions on data read from, or to generate data to be sent to the peripherals by higher-level commands. As an example, let us consider a disk controller: the processor may not control the disk in terms of input and output bits, but issues commands that are processed by the controller and converted into the low-level data exchange with the actual device. By analysing typical control applications, common I/O functions could be identified and provided as a standard library. These simple routines could be thoroughly (possibly formally) verified, contributing to the enhancement of system robustness. Their behaviour may be simple enough even to allow for safety-licensing. This feature alone would greatly enhance the safety of applications. Further, knowing exactly the functioning of the controlled peripheral devices, their parameters, their static and dynamic behaviour, and possible jeopardies, diagnostics could be very precise.

One of the simple functions to be migrated from task processors to intelligent peripheral interfaces is to poll the state of controlled system variables in order to synchronise controlled and controlling systems. Whenever possible, this technique is preferred to interrupts, because, first, the overhead of interrupt administration causes unpredictable delays and considerably reduces overall system performance, and second, state observation is much more fault-tolerant than event observation [72]. This was one of the basic ideas that motivated our work in this domain.

Further, the amount of data to be exchanged between task processors and peripheral interfaces through serial buses could possibly be reduced. It is most likely that the specific higher-level command itself will contain inherent information and will thus need to transfer less data than the detailed exchange between an actual peripheral device and its low-level interface. Sometimes, the same or very similar data sequences need to be transmitted. Further, there may, *e.g.*, be static pre-defined texts to be output in response to specific events. In the course of this, certain data can be pre-stored and programs pre-installed. When a task processor needs to send a message to the environment, only its index is communicated, but not the entire message.

Another possible function is recognition of pre-defined system states upon which certain actions need to be carried through. If the events result from state variable changes of the controlled system, relatively complex functions must be implemented in the environment to generate them. Sometimes, this may not be convenient or feasible. Instead, peripheral interfaces, which are observing the state variables anyway, may recognise such events and report signals to the operating system kernel processor as already described above.

3.4.2 Enhancing Fault Tolerance

Built-in local intelligence has an even more important effect, namely provision of enhanced fault tolerance. It is obvious that the best way to deal with faults is to prevent them from happening [9] in the first place. In this sense, the role of the measures described here is most important. If faults are detected and (possibly) dealt with at the lowest level, the rest of the system may not even be aware of them and, thus, does not need to react. As an example, pre-programmed functions can well support the fault tolerance of peripheral interfaces by self-diagnostics, self-check at initialisation time, and integrity checking during operation. In case of unsuccessful initialisation, the supervisory system can scan for alternative or redundant subunits. These units may be diverse, implemented with different technologies, have different performance, but still perform roughly the same functions and are controlled by the same commands.

If irregular behaviour is detected during operation, a system can be re-configured dynamically using redundant devices. Different units with similar functionality can be selected to replace the failed ones in a transparent way: the task processors are not aware who is performing the requested peripheral operations. If necessary, performance can be degraded gracefully, preserving essential integrity and safety.

Lately, instead of costly redundancy, reconfiguration of the resources remaining sound after incidents is finding more appreciation as another approach to assure fault tolerance. In a case of failure, the remaining intact components are re-arranged in a system with reduced performance, that may either continue to run the application at a lower-level mode, or be able to bring it into a safe state. Intelligence in peripheral interfaces is very useful in this sense; it may perform the reconfiguration locally without the need to disturb the rest of the system.

In the case of a non-recoverable error, or a failure of a task processor, the fail-safe property can be guaranteed by controlled and smooth driving of the controlled process into a safe state. This can be achieved by peripheral interfaces possessing enough intelligence to recognise faulty behaviour of task processors or other catastrophic states, and to put the devices they control into safe failing states.

To cope with transient system overloads when specified deadlines cannot be met due to exceptional irregular situations, system performance can also be gracefully degraded using peripheral interfaces. For instance, if a peripheral unit periodically outputting data to the environment does not receive the actual data on time, it can either repeat the last data item if this is adequate, or even predict the new one by extrapolation from preceding valid values. If the problems persist, the faults should not be tolerated any more; instead, a fail-safe shut-down should be performed.

In a similar way, an intelligent peripheral unit can deal with possible transient irregular output or input values. Since it is directly connected to the environment, it is in the best position to resolve such conflicts. Faulty inputs or outputs may be recognised based on the known physical properties of the controlled devices. The temperature of a large mass of a fluid, for instance, cannot change very fast; if the readings are such, a problem in data acquisition is recognised. Further, if a reading, following a row of regular ones, is totally out of logical bounds, it may be ignored as a transient error. In such a case, like above, the last reading may be repeated, or, time and performance permitting, a replacement calculated by extrapolation. Another example are unreasonable absolute values of physical entities (*e.g.*, temperature below $0\,K$).

3.4.3 Support for Programmed Temporal Functions

The intelligence added to peripheral interfaces can consistently support exact timing of input and output operations. This is very important for the stability of control systems and, thus, should comprehensively be supported by high-level programming languages and operating systems used in process control. In this subsection we show how temporal operations can be supported by intelligent peripheral interfaces.

One of the most serious problems in automation and regulation applications is jitter in both input and output process data. Due to the sequential nature of computer control systems it is not possible to read multi-variable input data or send corresponding output data simultaneously, causing the reactions of controlled systems to deviate from theoretical expectations. In our architecture the peripheral interfaces can be requested to perform I/O operations at exact time instants and, thus, in a fully simultaneous and synchronised fashion. This way, jitter can be eliminated at source.

The problem and its solution are presented in Figure 3.8. Sequentially reading process inputs may have the consequence that the values of the not yet read inputs change, and the state of the system is not acquired correctly. Similar for outputs: the set ones may already have effects before all are performed. This problem is solved if all inputs are read at the same instant, requested in advance by the processor, in a "sample-and-hold" manner. They are then read by the processor as the data are needed. On the output side, *vice versa*, data are output to the peripheral units, by which the actual output is performed simultaneously. The implementation is based on the fact that the peripheral units know the actual time. This information is maintained by the bus communication protocol (see Section 3.5). Correspondingly, commands issued by task processors include specifications of the time instants when operations are to be carried through.

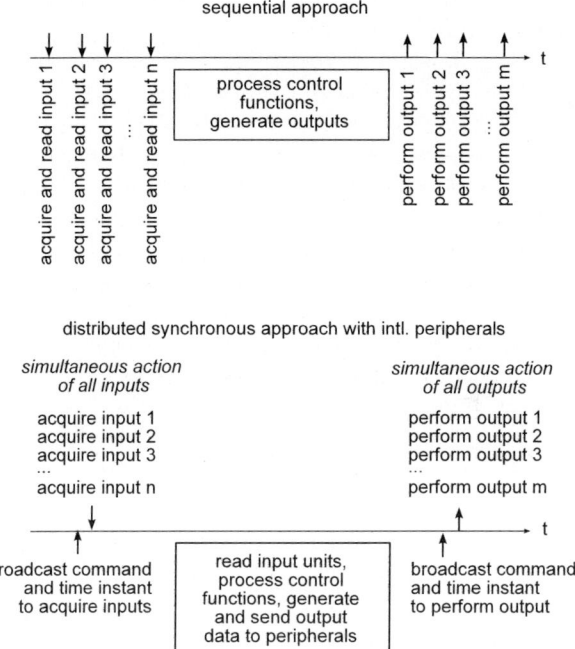

Fig. 3.8. Cause of jitter and its prevention by asynchronously operating intelligent peripheral interfaces

To produce an output at a desired instant, the task processor sends the corresponding clock value along with the output command well ahead of time. The peripheral interface addressed will actually perform the operation exactly at that instant. This would not be achievable by the task processor itself: due to task scheduling and various other delays, it is impossible to schedule any action in a task for a precise instant of time. Similarly, on input, a general processor can request an interface to read data at a certain future instant. Alternatively, input values can be acquired continuously and autonomously, time-stamped, and stored by a peripheral interface. Further, the data may be observed, and when certain anticipated events occur, only the current values can be time-stamped and stored to be read by the general processor later.

This solution is more general than, *e.g.*, the Sync and Freeze features Profibus [93]. These controls freeze the physical input and output data existing on one or more slaves simultaneously, like taking a snapshot. The selected slave(s) will stay in the frozen state until a cancellation is issued. In both Profibus and our system, the master has control over the exact instant of receiving or sending process data. However, in the case of Profibus, the master has explicitly to request the action at the very moment when it is to be performed, whereas our peripheral interfaces themselves administer their schedules.

Case Study: A/D Conversion

As an example, let us consider an implementation of an intelligent analogue-to-digital converter. In certain technologies, the time needed to convert analogue signals may depend on their values. Since the conversion times can, in general, not be neglected, special care must be taken to deal with them. One solution is to convert signals cyclicly, *i.e.*, A/D converters work continuously by triggering new conversions with their own "READY" signals. The results of the most recently completed conversions are stored with stamps of their acquisition times in a circular buffer, overwriting the oldest stored values. Given suitable conversion frequencies, the latest values can be considered valid, and can be reported to task processor, without any delay.

Before using this type of analogue data acquisition, it should be carefully checked whether the jitter introduced this way is harmful. If so, and if analogue data of specified time instants are required, more accurate values can be calculated from the stored data and their time-stamps by interpolation (or even extrapolation) carried out by the built-in processing facility.

Another possibility is to request a conversion at a certain instant. At this moment, the input value is sampled, then converted and stored until it is read by the task processor.

The digital value resulting from a conversion represents a certain physical quantity. Its true absolute value is related to the reading. It may be convenient if the peripheral interface, after conversion, also calculates the actual value according to a characteristic function given in an appropriate form. This value is then sent to the task processor as a response to its demand. Knowing the properties of the physical quantity measured, the results can be further evaluated. If any expectations are violated, corrective measures may be taken: the reading may be discarded, and a substitute value estimated.

3.4.4 Programming Peripheral Interfaces

In Chapter 4, an object-oriented high-level real-time language [114] will be presented in detail. Here, it is briefly mentioned how the language accommodates the access to peripheral interfaces. The code produced runs on the task processors. The language is not meant for programming peripheral interfaces themselves. As mentioned before, the peripheral interfaces are simple; their functions are specified, programmed and verified separately.

Using the paradigm of object-orientation, each type of peripheral interface can be considered a class, derived from a generic periphery class, and particular interfaces its instances. Their specifics, such as identification codes, must be described adequately (*e.g.*, by declarations). A simple inheritance scheme can be used to expand or modify characteristics of similar peripheral interfaces. Methods of the classes invoke pre-defined functions or procedures residing in the peripheral interfaces. They can either access the actual hardware registers, only, or perform certain functions. Remote procedure calls and/or parameter passing are system-dependent and physically implemented by system software. Apart from the methods, properties are specified in the classes, which are pseudo-variables and which can represent actual registers of the peripheral interfaces, or can implicitly be implemented by methods. Properties can be time-stamped, *i.e.*, methods to access them can be programmed to be activated at certain time instants.

3.5 Adequate Data Transfer

Data transfer in hardware architectures to be employed in embedded systems operating in hard real-time mode could introduce further sources of non-determinism. Generally, messages transmitted can be delayed by other messages, in some cases even indefinitely long. Together with the messages, task execution can also be delayed, jeopardising the feasibility of schedules and that deadlines are met. This was the rationale behind the evolution of data transfer capabilities as foreseen in the three stages of hardware architectures shown in Figures 3.5, 3.6 and 3.7, respectively.

In the first model (Figure 3.5), point-to-point communications between the kernel- and the task processors were used. In this way, messages from the different task processors could not interfere, and no delay could occur. Connections between the kernel- and the task processors were implemented using Inmos transputer links, *i.e.*, fast and simple serial connections. The choice was obvious, since the kernel processor was implemented using a transputer. Peripherals were connected to their respective task processors *via* serial local buses (at that time, Philips' I^{2}C or Motorola's M-Bus was employed.) Since the peripheral devices always worked in the slave mode, they could never initiate message transfers. Thus, the latter were controlled by the task processors, resulting in no collisions. The drawback of this architecture is a non-flexible structure with limited possibilities for reconfiguration because of the wired connections between the processors, and between peripherals and task processors.

In the second implementation (Figure 3.6), the point-to-point communication between the kernel- and task processors is retained, with its advantage of non-colliding data transfer. However, instead of being connected by local buses to the task processors, all peripherals reside together on the common local bus driven by the kernel processor, so that

- Events from sources directly wired to the kernel processor can now be transmitted by messages on the common local bus.
- Task processors may now share use of the peripherals, allowing for better flexibility and even for a homogenous multiprocessor structure, although adequate mapping and scheduling in such systems are serious problems.
- Fault tolerance is improved by redundant peripherals that may also be shared among the task processors.

Since the peripherals still behave as slaves, it is the exclusive responsibility of the kernel processor to manage traffic, thus avoiding collisions.

In the third case (Figure 3.7), all units reside at the same local bus. It is now used for both communication between kernel- and task processors and for peripheral interfaces access. There is a danger of collisions that may lead to non-predictable delays. To prevent it, a time-triggered protocol is employed, which is detailed below.

3.5.1 Real-time Communication

In distributed control systems, a set of processing elements, sensors, actuators, and plant components is interconnected through some means of communication. There is a number of implementations of such solutions already used in real-world control applications (*e.g.*, Controller Area Network – CAN). For fault-tolerant control systems, even during maximum busload, timely transmission of all safety-related messages must be guaranteed. Occasionally, several messages may be produced to be transmitted at the same time. Hence, some messages will be delayed.

Traditionally, message collisions on a bus are resolved using priority-based algorithms: the message with the highest priority assigned is transmitted first. Such an approach is employed in the widely used CAN protocol. The main problem with priority-based protocols is the impossibility to predict and guarantee worst-case transmission times. Because of dynamic message arrival and transfer times, it may happen that a low-priority message is blocked indefinitely long by higher-priority ones. As a last resort, some sort of a bus guardian must be used to prevent too frequent high-priority messages (possibly as consequence of an error) blocking the low-priority ones. Protocols with random collision resolution (like standard Ethernet) are not sensitive to wrong priority assignment. However, the danger of long and non-deterministic blocking of messages still remains, and is only depending on chance.

In the last five years, a lot of effort has been devoted to introduce real-time capabilities into Ethernet. [39] Convenience of Ethernet networks, existence of good infrastructure and devices, and the widespread TCP/IP protocol let it appear attractive to employ Ethernet also in industrial process control environments. There already exist more than ten different solutions, *e.g.*, Ether-CAT, ModBus/TCP, Ethernet-Powerlink, SERCOS III, PROFINET, ETHERNET/IP, or VARAN. As an example, let us consider a protocol proposed by Hoang, Jonsson, Hagström, and Kallerdahl [55], providing enhancements to full-duplex switched Ethernet in order to be able to give throughput and delay guarantees. The switch and the end-nodes control real-time traffic by earliest-deadline-first scheduling on the frame level. In the proposed solution, there are no modifications to the Ethernet network interface card, which allows the real-time nodes to be interconnected with existing Ethernet networks. Although there are standardisation efforts already on their way, at the moment, it looks like that there will be one standard document IEC 61784-2 specifying at least ten different and mostly incompatible technical solutions for Real-Time Ethernet. This situation is unfortunate to the user, who would like to have a single, adequate and usable real-time Ethernet protocol.

3.5.2 Time-triggered Communication

Recently, the so-called time-triggered approach is becoming increasingly popular. In the aerospace industry, time-triggered applications are state-of-the-art already today. Time-triggered automotive applications are becoming part of the next product generation. By this approach, message transfers are scheduled at design time instead of at run-time. In general, each message gets its unique and statically defined access slot within pre-defined and globally synchronised time-frames. Systems run with pre-determined timetables.

There are several competing time-triggered solutions to control buses. A number of them are based on the CAN bus protocol as network data layer because of its simple and inexpensive infrastructure. One of such approaches is Time-Triggered CAN (TTCAN) [64]. A good description, which the part below is based on, is given in [43].

In TTCAN, one of the nodes on the CAN bus is designated as the *time-master*. It periodically generates administrative *reference messages* marking the beginnings of the *basic cycles*; see Figure 3.9. The nodes that need to exchange messages detect the reference message and wait for their designated *time window*, within which they send or receive a message. Since broadcasting the reference messages is under the control of the time-master, it represents a potential risk. If this node fails, communication stops representing a central point of system failure. To solve this problem, time-triggered protocols allow for several potential time-master nodes to co-exist in a system. One of them is the current time-master that starts sending the reference messages at system start-up. If the current time-master fails, another time-master becomes the current one after arbitration. The role of time-master can be combined with the role of other entities that may represent a single point of failure. For example, the node acting as time-master may also house the kernel processor in a non-distributed configuration. Thus, if the time-master node fails, its role is transferred to another one together with the functionality as kernel processor.

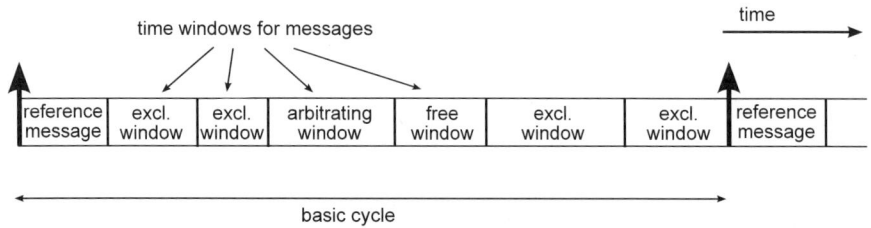

Fig. 3.9. Basic cycle of TTCAN

There are three different kinds of time windows:

Exclusive Time Windows within which periodic messages between nodes are transmitted as scheduled in the pre-determined timetables,

Arbitrating Time Windows within which sporadic messages and, possibly, competing soft real-time messages are transmitted after arbitration using a specifically defined protocol, and

Free Time Windows for future extensions.

Different messages may require to be transferred more or less frequently. For that reason, several different basic cycles with the same distribution and duration of *transmission columns* may be bundled together in sequence; see Figure 3.10.

The sequence of basic cycles and of transmission columns for a specific application is defined in a *system matrix*. It is composed at a design time based on transmission needs of communicating partners. Its composition must comply with the requirements, and should at the same time maximise the number and the length of arbitration- and free slots. In complex cases this task is not trivial, and may even require certain artificial intelligence approaches (see, *e.g.*, [115, 40]). From the system matrix, the information for triggering the data transfers is acquired. For instance, *Tx_Trigger* starts transfer of a message. It consists of the transmission column's number, window type, cycle offset (if there are several different basic cycles in the sequence), and of the repeat factor, which counts how many times in the bundle the message is transmitted.

Any TTCAN controller maintains its local time counter, which is incremented each Network Time Unit. (NTU) Another timer is necessary to achieve

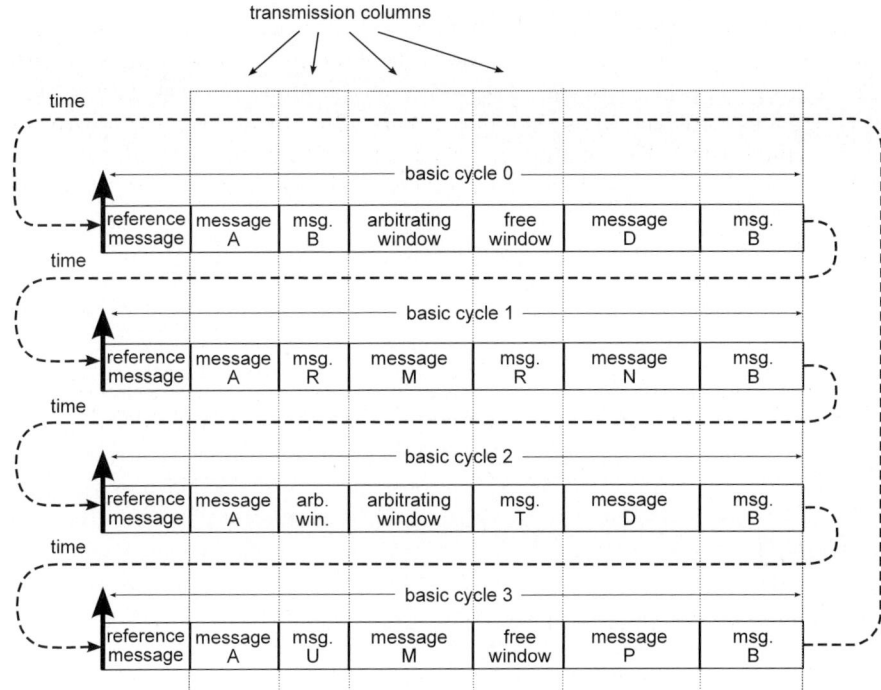

Fig. 3.10. Sequence of different basic cycles

cycle-based sending/receiving of the messages and to synchronise to the cycle time of the time-master in a TTCAN network. According to the synchronised timers, specific transmit/receive triggers generate the actual transfers. Reference messages are not always issued periodically. In some applications it can be advantageous to trigger the transmission of a reference message by an event external to the bus; see Figure 3.11. In this case, the application has to signal to all other bus members that the next reference message will be delayed. This is done in the reference message preceding this event synchronisation by setting a specific control bit, the *Next_is_Gap* bit. In order to avoid non-deterministic delays, for any application the maximum length of the gap must be defined. The next reference message is then postponed until the external event occurs, or for the time gap, whichever occurs first.

There are two levels of extension in TTCAN. At level 1, a time-triggered functionality is guaranteed by reference messages of the single time-master. In this model, the existing CAN infrastructure can be used. Only the high-level protocols must be adapted, what can be done in software. For fault tolerance, several redundant time-masters can be employed. At extension level 2, a globally synchronised time-base is established, and a continuous drift correction among the CAN controllers is implemented. This extension requires specific hardware devices. A similar approach is used by the Time-Triggered Proto-

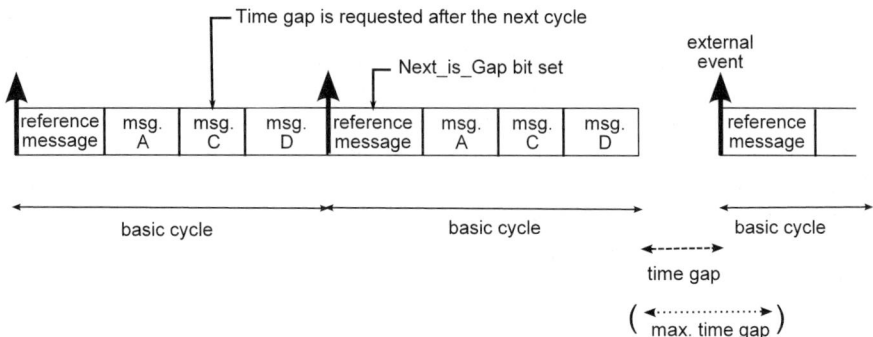

Fig. 3.11. Synchronisation of the basic cycle with external event

col (TTP and TTP/C) [45], which can achieve much higher data rates than TTCAN. Compatibility with CAN is implemented at several levels. Apart from the CAN-TTP-Gateway solutions, a CAN emulation layer can be used on top of TTP: the registers of a CAN controller module are emulated allowing to re-use existing software. As a drawback, the TTP solutions with higher data rates require specialised hardware. Other solutions try to combine the best properties of both the priority-based and the time-triggered approach. One of such protocols is FlexRay [5], which permits the co-existence of both prioritised and time-triggered messages on the same network. At this time, it is still not sure which approach and which solutions will prevail in the future.

3.5.3 Fault Tolerance in Communication

When fault-tolerant behaviour is required, other issues emerge. A single communication bus represents a single point of failure in a system; thus some kind of communication redundancy is needed. As an obvious solution, triple-modular redundancy, a fault-tolerance measure widely used in other domains, would mean that buses should be triplicated in order to detect faulty behaviour of one. Triple- or n-modular redundancy, however, has disadvantages, the most obvious being cost. One purpose of redundancy, *viz.*, to detect the faulty replica, could in some cases also be accomplished with other means. Instead of redundant devices, a simpler mechanism based on monitoring can be used to detect faults. Moreover, usually the existing bus control protocols already include some sort of error detection that can easily be improved with additional dedicated hardware.

Instead of using three or more buses, a reasonable degree of fault-tolerant behaviour could be achieved with only two. Since it can be expected that a system will operate without a failure most of the time, it would be a waste of resources if secondary buses would be used only as back-ups. Thus, as long as no error is detected by the monitoring system, both buses should be used for data transfer. When, however, a fault is detected on one bus, communication is switched to the other one. The bus remaining operational must handle all necessary traffic, perhaps by gracefully degrading the performance, and/or eliminating some non-critical messages altogether.

There are several possible scenarios regarding how to utilise multiple communication channels. The most obvious is load-balancing with the bus-load dynamically equally shared among them. This way, shorter transmission and reaction times can be achieved. Dynamic load-sharing, however, increases complexity and is, thus, less safe. In simpler cases, different communication scenarios can be prepared in advance. When there are no faults, certain messages can be mapped onto certain buses. For example, one bus is dedicated to the safety-critical messages (by using fault-tolerant protocols with lower throughput), and the other to less important messages (using, *e.g.*, priority-based scheduling). Another example is to transmit system messages on one bus and application messages on the other.

In case of faults, communication channels can easily be reconfigured if messages are prioritised. Then the frequencies of the less important messages can be reduced, or these messages can be eliminated completely. For time-triggered operation, when a bus fails a new time schedule must be applied to the remaining bus(ses). There must be a set of different timetables prepared in advance for each failure mode. The numbers of possible timetables exponentially increase with the number of redundant buses, and each combination must be elaborated in advance at design time. Another difficulty of time-triggered operation on several buses represents the global synchronisation of the time-frames.

Not only the faults on the communication channels need to be considered in a fault-tolerant communication system. Faults in processing units, sensors and actuators can have important impact and must be handled by the communication infrastructure. For example, if an actuator fails, the messages sent to it may be re-routed to a redundant one. This can be implemented by considering all redundant resources of the same type as a single device. Each message for the actuator is sent to all operational replicas, but only one (or, in certain cases, all) of them are then responsible to perform a certain action. On the other hand, each replica of the same sensorial resource can provide a message to the system at the same time. Each receiver then decides which message is used for further computation. In this case, it is important that a sending device can automatically detect if it operates properly. It must send an appropriate status (*e.g.*, invalid value; fail-safe approach) or remain silent (fail-silent approach). Of course, this sort of redundancy decreases the usable throughput of a bus. However, it also reduces the need for additional buses and arbitration hardware. If a processing unit fails, the whole system must be reconfigured and the message channels may be re-routed, so that they can be served by a spare unit.

When custom-built components are used, fault-detection can be built in, and performed by the components themselves by means of appropriate hardware and software. However, in the case of the Commercial Off-The-Shelf

(COTS) and special plant components, this is usually not possible. Then flexible monitoring systems should be integrated into communication subsystems to monitor the signal- and data-flows between components, sensors and actuators. In simple situations, monitors can detect out-of-range values. In more complex implementations, they can monitor the dynamics of the signals, the mapping between inputs and outputs, temporal restrictions, *etc.* Such systems also monitor the behaviour of the buses and are able to switch between them. In Figure 3.12, a possible implementation of such a gateway between non-fault-tolerant communication protocols and/or buses and fault-tolerant ones is given. Further, it is shown that not always all devices need to be connected to both buses: in case of failure a certain functionality may be renounced, like a sensor, an actuator, or even a processing unit (*e.g.*, S2 and PU2 in the figure). If the bus is not operational any more, after reconfiguration the units are simply ignored.

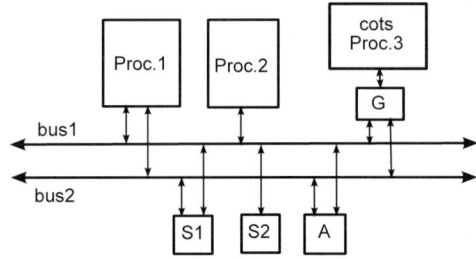

Proc.$_x$ – processing unit, S_x – sensor, A – actuator, G – NFT-FT gateway

Fig. 3.12. Employing a COTS processing unit in a FT bus architecture

3.5.4 Distributed Data Access: Distributed Replicated Shared Memory

In process control applications, control functions are implemented in the form of function blocks that communicate with each other, read data from sensors, and produce signals for the actuators. Function blocks are generalisations of control algorithms in the form of programs. They may be implemented as tasks in asynchronous multiprogramming environments, or just as subprograms or routines. In distributed control applications, both function blocks and peripherals are mapped onto the hardware architectures. To access data it is necessary to know on which processing node a specific function block resides, or where and how to address a specific I/O device.

When designing a control application on a distributed control system, the notions of nodes, messages, timetables, *etc.* are usually not transparent from the application designer's point of view, who is dealing with logical issues, such as tasks and variables. In order to relieve him or her from application details, mapping between views is helpful. To allow for this, in this section a specific approach called distributed shared memory model is introduced to isolate control applications from the issues of hardware implementation.

To achieve this isolation, the underlying communication system is only exposed as a set of data containers called memory cells. Two processes use the same cells to communicate with each other regardless of their locations (*i.e.*, on which processing nodes the processes actually run, on the same or on different nodes). The same mechanism is used for the communication between control applications and sensors or actuators. Thus, shared memory cells serve as sources or sinks of the virtual communication links between application, hardware platform, and environment.

An example is shown in Figure 3.13. The application logically consists of three function blocks, two process inputs and two outputs, which all communicate with each other. For instance, FB2 receives information from Input1: the latter writes information into memory cell MC4, where it is available to FB2. FB2 and FB3 communicate with each other in both directions using MC3 and MC7. They both send data to FB1 *via* MC2 and MC6, respectively. In turn, FB1 controls Outputs1 and 2 *via* memory cells MC1 and MC5.

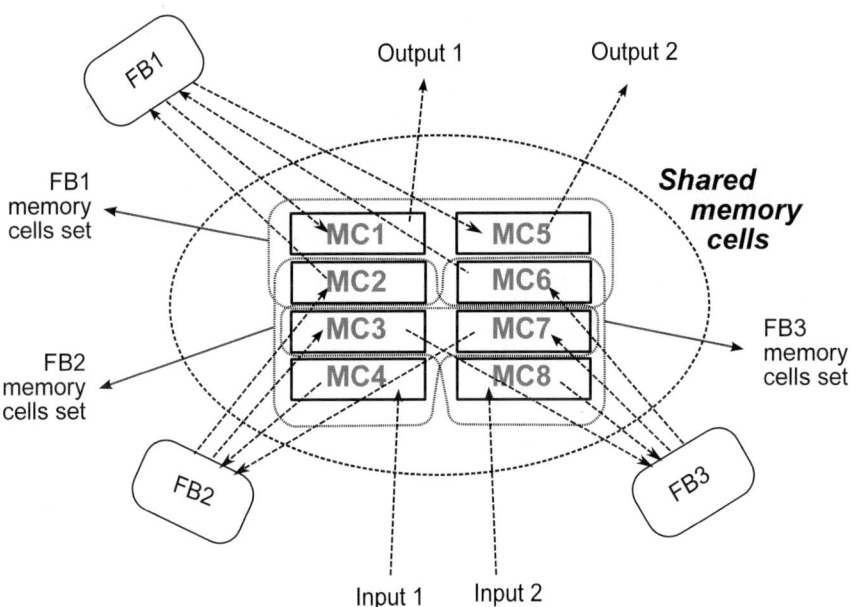

Fig. 3.13. Model of data transfer among function blocks and peripherals

Partners in communication access their memory cells in their usual memory address space. If communicating units reside on the same node, the same cells in the form of memory locations are shared among them. If, however, they are running on different nodes, data written to a cell are transparently distributed through the system by means of TTCAN messages. Each communicating node accesses its own set of memory cells, which in Figure 3.13 is represented by a dashed rounded rectangle. FB1, for example, only needs to "see" memory cells MC1, 2, 5, and 6, Input1 can only access MC4, *etc.* Cross-sections of rectangles represent communication cells.

Process data acquisition is a sensitive part of process control. In order to prevent faults in further processing, it is reasonable to detect faulty input data at the very beginning. For that purpose, each data cell can be associated with a validation routine, which may perform different kinds of testing and validation of data. The same routine can be used for mapping between different data domains used by the application and the peripherals. Because data are indirectly transferred between control applications and peripherals, desired data transformations can be applied. Control applications usually deal with abstract representations of certain physical quantities from the environment (*e.g.*, temperature represented in Kelvin). On the other hand, acquisition and actuation of these quantities are performed by simple input/output devices (*e.g.*, A/D converter connected to a temperature sensor). These devices use their own kinds of data representation. To simplify the development of control applications, system software can transform between different value domains in transparent ways; see Figure 3.14.

In the example, voltage provided by a temperature sensor is first converted into an integer and then transformed into a floating-point value appropriate to the application. To allow for this, a memory cell is associated with two transformation functions. The first is applied when data are written to, and the second is applied when data are read from the cell. If, for some reason, the temperature sensor must be replaced with one of different type, only the transformation functions need to be updated accordingly. These functions may be implemented either in software or in firmware of ASIC – programmable hardware devices (FPGA, SOC, *etc.*) The latter option is less flexible, but safer.

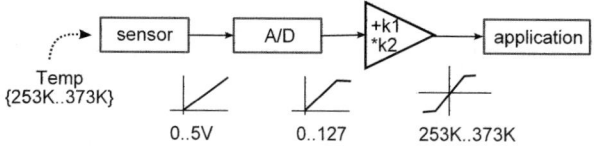

Fig. 3.14. Mapping of domains between the A/D converter and application-related floating-point units

In addition to actual application data, memory cells may also contain some other attributes that can be provided by the applications (*e.g.*, quality of information generated by a producer) or the systems software (*e.g.*, timestamp when data has been generated). System software can also monitor when data in cells actually change, and signal this to control application, *etc.*

When both source and destination peers for communication reside on the same processing node, simple memory transfer of data within common global memory can be used. On the other hand, when communication spans over two or more nodes, the data written into a cell in one node must be transparently distributed to all appropriate nodes in the system by means of some sort of transparent data transfer in the background.

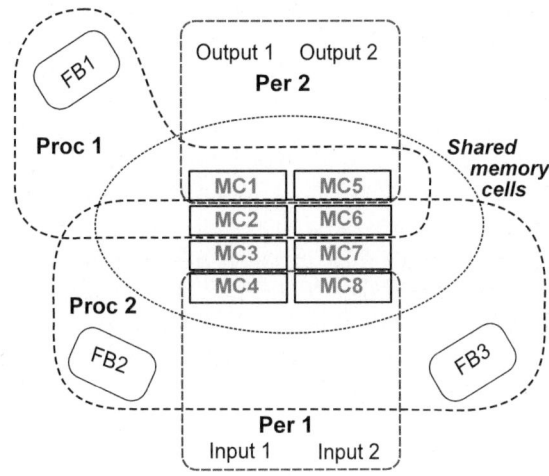

Fig. 3.15. Mapping of function blocks and I/O to processors and peripheral units

In Figure 3.15, following the above example, function block FB1 is running on processor Proc1, while FB2 and FB3 are mapped onto Proc2. Inputs1 and 2 come from peripheral unit Per1, Outputs1 and 2 reside on Per2. While FB2 and FB3 can communicate *via* MC3 and MC7 using common global memory, for the data transfer between FB1 and FB2 *via* MC2, and between FB1 and FB3 *via* MC6, a communication mechanism is necessary. The same holds for the transfer of all I/O data.

The implementation architecture of the approach of distributed replicated shared memory cells is depicted in Figure 3.16. Function blocks running on a certain processor only need access to a certain subset of memory cells. Each processor keeps its own replica which is periodically updated among partners sharing the same cells. Application software running on a processor accesses its subset of memory cells residing in its local memory. To allow for virtually simultaneous access from application software, and updating from the distributed shared memory system, local memory is implemented in a dual-port technology. Hardware support takes care of rare conflict situations, when requests from both sides arrive simultaneously. Local memory cells are cyclically updated with other replicas by underlying firmware. The values are taken from the messages periodically transmitted over a bus. These messages are generated by the producers of the information for the relevant memory cells.

As the means of communication, the TTCAN protocol mentioned earlier is most suitable. Contents of the memory cells are generated by producers and contained in messages. If the contents of a cell changes, the value of the memory cell is overwritten on all nodes on which it is present by a firmware-implemented mechanism. It runs transparently on the processing and peripheral units in a system, connected by the TTCAN bus. The mechanism is triggered (and thus synchronised) by the basic cycle's reference messages.

Fault Tolerance in Data Transfer

Apart from the fault tolerance offered by the TTCAN protocol, as mentioned in Section 3.5.3, the proposed transfer mechanism of distributed replicated shared memory inherently contains further important features, enabling easy reconfiguration.

An example is shown in Figure 3.17. For better overview, the memory cells are not shown and are supposed to be included in the function blocks. In normal fault-free mode, Driver1 from the peripheral unit reads Sensor1 values, possibly doing some processing on them, checking their bounds and, finally, generating messages, in synchrony with the TTCAN schedule and timing. FB2 running on Proc.2 takes in this information, does some further calculations, and sends the results over the message M11 to FB1 on Proc.1. Data-flow is shown by dashed arrows. Assuming that Proc.2 is more error-prone than Proc.1, the designer may consider preparing an alternative function block, FB2', which can, to a certain extent, replace FB2 in case of failure of the latter. It runs on the more robust, but less powerful Proc.1. Thus, it should

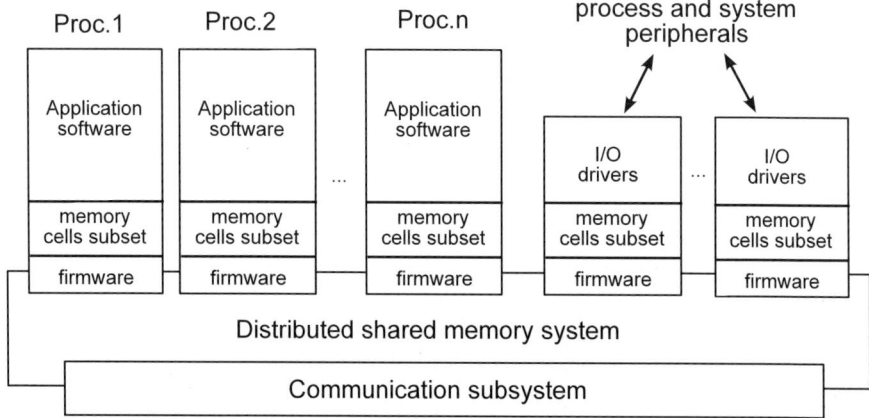

Fig. 3.16. Architecture of the distributed shared memory approach

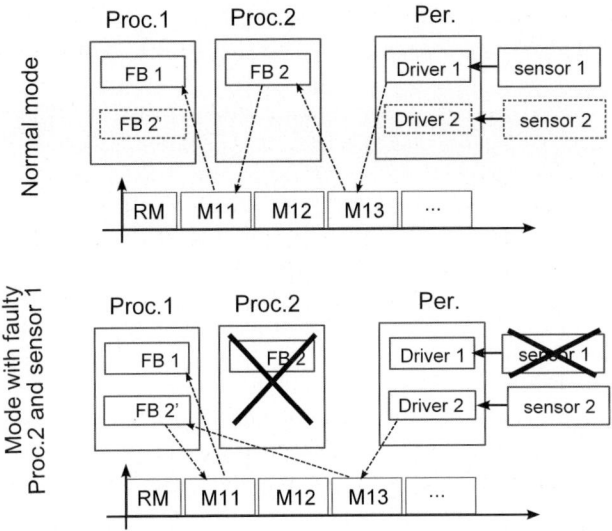

Fig. 3.17. Reconfiguration in the case of a unit failure

be supposed that the performance of FB2' is more modest, and the functionality must be degraded gracefully. The function block is pre-loaded on Proc.1. A further assumption is that Sensor1 may also fail, and back-up Sensor2 is prepared to replace it. In case of failure of Proc.2 or/and Sensor1, the system is re-configured. Instead of the faulty Sensor1, Sensor2 will generate input values. These are, instead of the faulty FB2, now read by FB2'. This function block also supplies the data to FB1.

By using the shared memory model, it is also possible to perform a simple and unified kind of tracing, logging, and diagnostics of important state variables used by control applications. Because information written to a cell is transparently distributed to all system nodes, it is possible to have a dedicated monitoring node that acquires the current system state, saves it and, possibly, evaluates its plausibility.

4

Programming of Embedded Systems

Similar to the development in other domains of embedded systems, in practice control applications are often programmed by improper means. For a number of pragmatic reasons, the same methods, techniques, and programming languages as for general-purpose desktop applications are employed. One of the reasons for this is common knowledge of programming: the languages and methods learned can do the job to a certain extent, so most practitioners use them with little concern about the rather specific circumstances of embedded systems.

There is, however, a number of aspects of building embedded computer systems that differentiate them from the development of traditional computer applications. In addition to the requirements that can be found there, embedded systems necessitate many kinds of interdisciplinary requirements to be fulfilled. Below, some characteristic groups of them are enumerated. The most important issues will be discussed in more detail later.

Consideration of time. There are several inherent temporal characteristics of control systems. One is reaction time, *i.e.*, the time interval between the moment when a particular condition occurs, and the moment when the control system starts to react upon it. Another characteristic is task execution time, *i.e.*, the time needed to perform a specific task. It depends on the capabilities of the hardware platform and the system software of the control system (such as speed of the microprocessor, throughput of the communication infrastructure, or latency of tasking operations).

Based on these characteristics, two parameters are usually set and included into the system requirements for each task in a control system: its deadline and its execution time. The *deadline* is a property of the system as a whole and must be met — at least for hard real-time tasks — even in worst-case scenarios. Whether or not the deadline will be met depends on the time needed for the system to detect the task triggering conditions, the time the system software needs to react on it, and the *execution time*, *i.e.*, the time the control application spends performing the appropriate actions. In addition, of course, the load all other tasks impose on the system must be considered. The execution time of a task is important for the scheduling algorithm employed and for schedulability analysis. Usually, only the worst-case execution time is relevant.

Another temporal characteristic of a control system is how frequently a certain task-triggering condition will occur in the system. This characteristic is important for schedulability analysis. For example, an air-conditioning system will probably have a relatively long time between two successive requests for changes of the temperature regulated. In contrast, for a radar tracking system, the target's position may change very frequently. It is also important to examine the distribution of event occurrence, *i.e.*, whether events occur periodically or sporadically. Periodic events with long intervals are much easier to deal with by control systems than sporadic ones with high frequencies of occurrences.

Consideration of other non-functional requirements. Usually, there are some other requirements specific to the application domain set for a control application that cannot always be expressed in the program code. These requirements are commonly also safety-related features of the system and must be dealt with consistently. For example, in a radiation device for cancer treatment, it must be assured that a maximum dosage of radiation will not be exceeded; or for an elevator, the maximum acceleration allowed is limited.

A control system usually has no provision to measure directly such quantities; even if it has, there must exist some other means to assure that safety is retained even in the case that the control system fails. Coming back to the cancer-treatment example, there must exist an independent device that measures the current level of ionisation exposure and raises alert in case of a possible overdose. Similarly, the elevator must have a mechanical emergency braking system to be engaged if the acceleration is too high.

Dynamic system behaviour and distributed computing. Almost all applications of embedded systems consist of certain numbers of autonomous assignments that must be executed in parallel, communicate with each other, and synchronise their execution. In distributed control systems, these assignments may be executed on different processing units. All this requires explicit support for modeling and programming parallel activities.

Diverse target platforms. Desktop computer applications are currently being executed on only around a dozen hardware platforms. Moreover, there are just a couple of major processor vendors for this kind of platforms. On the other hand, there exist a vast number of different platforms and processors for embedded systems, and also a variety of input/output devices for them. For these reasons, it is very difficult to build a universal embedded control application that can be used on various kinds of targets. Usually, software libraries for the platforms are missing, and programmers must deal with each individual component of hardware separately. Therefore, it is wise to separate the algorithmic part of an application from the device driver routines. Some of the operating systems for embedded systems include a so-called Hardware Abstraction Layer. This additional layer virtualises the use of input/output devices and microprocessor platforms, however at the expense of efficiency and speed.

Limited resources. There is high pressure on the producers of embedded systems to reduce the expenses of building controlled systems to a minimum. Usually, sacrifices are necessary with respect to the amount of available memory and others hardware resources, *e.g.*, to achieve low power consumption. A lower-performance microprocessor causes longer execution times and, therefore, increases the likelihood of system overload or deadline misses. Lower memory capacity limits the size of program code and data structures. These shortcomings introduce another level of complexity to system designers and programmers, *viz.*, how to find solutions complying with the limited assets. Code must often be optimised by hand, the complexity of algorithms must be reduced, or the accuracy of results must be degraded. If power consumption is an issue, another kind of measure must be taken that puts the hardware components currently not in use into stand-by mode. There is little support to deal with these circumstances. A good compiler can effectively optimise the amount of space used; usually, however, there is a trade-off with execution time. Moreover, virtually no tools available can help if the power supply is limited. Therefore, all this requires good knowledge of the components used and experienced programmers.

Low-level access to hardware. A typical control system interacts with the systems controlled and the environment through various sets of sensors and actuators. In contrast to typical desktop applications, which work with a well-known and limited set of I/O devices, different and usually solution-specific hardware components are required. To use them properly, in-depth knowledge of them is usually required allowing to influence them on low levels.

Fault tolerance and safety criticality. The vast majority of control systems must be considered safety critical, *i.e.*, a fault in (at least a part of) an application can lead to situations where relatively large material losses may occur or even human safety may be endangered. Another important aspect is in many cases the availability required of a control system, such that production does not stop even if there is a fault. Therefore, embedded systems must be robust enough and must react to faults both in the plants controlled and in their own hardware and software in a controlled manner. Here, distributed architectures of embedded systems guarantee more flexibility in dealing with faults. If a component fails, the functionalities lost can be replaced by redundant resources from other nodes, possibly with no degradation in system performance.

Long exploitation time. Embedded systems are intended for use during long times without upgrades and with as little maintenance as possible. A large number of control systems are physically implanted into controlled devices, or they are placed in isolated and hazardous environment where they are not easily accessible. Another reason for long exploitation is the expensive development. To build a dependable control system and to prove formally that it will work within specifications all the time is time-consuming and costly. When something is changed, the validation and certification processes must be repeated. As an example, the control systems of the Space Shuttle are almost the same as were already used in the mid-1970s [111]. The Space Shuttle has APA-101S computers (five of them for redundancy) running at about 1.2 MIPS, and equipped with a few megabytes of ferrite core memory featuring resistance to radiation. The entire control software for the Shuttle is less than 1 MB in size.

On the other hand, software upgrades are in some cases not so uncommon anymore. It has been observed, for instance, that the ignition control software in certain cars has been upgraded for better fuel consumption and, thus, higher energy efficiency. The following example illustrates that software upgrades can be used to overcome flaws in the hardware. When the main antenna of the Galileo spacecraft did not deploy fully, the mission was in jeopardy [14]. There was no possibility for any repairs. With software upgrades providing better data compression algorithms, however, it was possible to regain many of its capabilities.

Holistic approach to develop all system components. In developing stand-alone computer systems, developers usually focus on the software using some off-the-shelf operating system, supporting libraries, a commercial database, *etc.* For embedded systems, however, it is usually necessary to develop not only the software but also the hardware, and sometimes even the operating system. Advances in technology mean that parts of the software algorithms can actually be implemented in hardware (*e.g.*, by means of field programmable gate arrays). To obtain satisfactory results, all components of such control systems must be designed in a holistic way. This methodology is also essential if temporal predictability and dependability are decisive issues.

Human factors. Application developers of embedded systems are usually control, electrical or mechanical engineers. They have a broad knowledge in their problem domains, are skilled in designing control applications, but often have limited knowledge of computer engineering and computer science. Since they are usually swamped with work and do not have much time to learn further tools and methods, the means offered to them for application design should be problem-oriented, user-friendly, and not too complicated to learn. The design results must be easily understandable, surveyable, and verifiable with reasonable effort.

This is probably also the reason why formal methods and sophisticated abstract design approaches are not widely accepted in these communities. Instead, strict specification and verification should be integrated into the design process in a problem-oriented way, without imposing too much additional work on often already overloaded application designers.

4.1 Properties Desired of Control Systems Development

Based on the particularities discussed in the previous section, some elements can be identified that should be part of a programming infrastructure for embedded control systems. Most of the commercially available development tools lack, however, at least some of them. To cope with this situation, programmers can apply sets of formal organisation rules that mimic the desired properties indirectly. For example, one of these features is explicit task declaration. Instead of being a construct inherent to the programming language, most development environments provide task creation by system software calls, only. By declaring appropriate macros and/or routine wrappers, the particularities of system calls can be concealed. If code is ported to a platform with a different operating system, only the wrappers must be modified. Some development tools even allow implementation of custom-made development environments through a set of well-defined application interfaces. Some features that greatly enhance the suitability of a programming language for embedded applications will be discussed below.

4.1.1 Support for Time and Timing Operations

It is paradoxical that the notion of time, being the most characteristic feature of real-time systems, is largely neglected in most programming languages for embedded applications. Development tools and programming languages should explicitly include support for different presentations of time and operators to be used with temporal data types. Moreover, the temporal behaviour of the applications themselves (execution time, deadlines, *etc.*) should also be definable and observable.

The majority of programming environments treats time as integer or floating-point values. For example, the data type `clock_t` used in the time.h library of the ANSI C language is actually an integer. The sizes and precisions of the data types representing temporal quantities should depend on the types of applications. A control system with short reaction times, for instance, requires very detailed and accurate representations, while, on the other hand, in some slower processes the time quantum can be larger.

Temporal Data Types

There are two different aspects of temporal values. The first one denotes a specific, absolute instant (point in time, *e.g.*, "at noon" or "8:00 hours"), and the second one represents a relative time passed, a duration (time interval, *e.g.*, "5 s"). For this purpose, some programming languages (the most characteristic example is PEARL [29]) introduce two data types, CLOCK and DURATION, which represent these two aspects of time, respectively. Further, a set of well-defined arithmetical operations is defined for the temporal data types, as listed in Table 4.1. For example, to compute the duration between two absolute points in time, their values are simply subtracted. A time interval is always a non-negative quantity.

Table 4.1. Result types of arithmetical operations on temporal data

```
numeric * duration = duration
duration / numeric = duration
duration + duration = duration
duration - duration = duration
clock + duration = clock
duration + clock = clock
clock - duration = clock
clock - clock = duration
```

At the implementation level, temporal quantities must be represented by generic data types and, finally, mapped onto physical memory. Relative time intervals are typically represented as integers expressing a number of basic time units (*e.g.*, milliseconds). However, for programming convenience, other time units can be used within the applications, which are then internally converted into the basic ones. For absolute time, representations are more complicated and possible in different ways. One is to pack absolute time into a structure comprising different time parts as is shown in Figure 4.1.

```
absolute_time struct {
   int year,
   int month,
   int day,
   int hour,
   int minute,
   int second,
   int millisecond
   }
```

Fig. 4.1. Structural representation of absolute time

On one hand, this representation facilitates the conversion of the internal representation to a form understandable by humans and *vice-versa*, whereas on the other hand it renders the implementation of temporal operations very difficult. For example, if a variable of this type is incremented by 1 min, a lot of testing and calculation must be performed. So, if after incrementing the number of minutes exceeds 59, the hours must also be increased, then perhaps the days, *etc.* The system software must also consider the number of days in months and leap years. In addition, to increase the precision of temporal variables, further elements must be added to the structure.

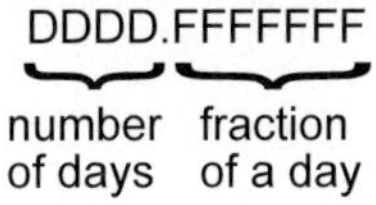

$$\underbrace{\text{DDDD}}_{\substack{\text{number} \\ \text{of days}}}.\underbrace{\text{FFFFFFF}}_{\substack{\text{fraction} \\ \text{of a day}}}$$

Fig. 4.2. Representation of absolute temporal values with Julian day numbers

Another form of representing temporal values is to use floating-point numbers in a form resembling the Julian date (Figure 4.2). The Julian day or Julian day number (JDN) is the integer number of days that have elapsed since noon of the Julian calendar's day 0, *viz.*, Monday, 1 January 4713 BC. Negative values can also be used. As an example, noon of 3 April 2007 carries the Julian Day Number (JDN) 2454193 and is divisible by 7, because it was also a Monday. The fractional part of the JDN gives the time of day as a decimal fraction of one day, with 0.5 representing midnight. Typically, a 64-bit floating point (double precision) variable can represent an epoch expressed as Julian date to about 1 ms precision. For details about fractional values see Table 4.2. For computer implementation and use in technical systems, the traditional JDN is usually slightly adapted in order to save memory space. The time origin is thus changed, *e.g.*, to 1 January 1900 in Linux. Also, the fraction usually starts with midnight instead of noon for practical reasons.

Table 4.2. The decimal parts of a Julian date

$$0.1 = 2.4 \text{ h or } 144 \text{ min or } 8640 \text{ s}$$
$$0.01 = 0.24 \text{ h or } 14.4 \text{ min or } 864 \text{ s}$$
$$0.001 = 0.024 \text{ h or } 1.44 \text{ min or } 86.4 \text{ s}$$
$$0.0001 = 0.0024 \text{ h or } 0.144 \text{ min or } 8.64 \text{ s}$$
$$0.00001 = 0.00024 \text{ h or } 0.0144 \text{ min or } 0.864 \text{ s}$$

With this representation, the transformation of time into a form readable to humans becomes more complicated. It does not present a problem, however, since this is carried out by a compiler when programming and by a user interface at run-time. Temporal arithmetic is highly facilitated as basic floating-point operations are employed. For example, to increase a time variable by 1 min the value $\frac{1}{24\times60} = 6.94\dot{4} \times 10^{-4}$ must be added.

The granularity of temporal values depends on the precision of floating-point numbers. Single precision floating-point numbers yield about seven decimal places. That means that the granularity is up to $\frac{1}{100}$ of a day (five decimal places are needed to represent the number of days). This is more than 14 min and, therefore, not suitable for most cases. Double precision floating-point numbers already provide 15 decimal places with a granularity of less than $1\mu s$. This is probably sufficient for most applications.

A drawback of this representation is that it is not suitable for simple processing elements without floating-point support. Once a quantisation of time (*e.g.*, 1 ms) is defined meeting the time constants of a process, however, elapsing basic time units can be counted in integer variables. In other words, the JDN is multiplied by a factor such that only the integer part remains. The absolute time can then be interpreted as number of basic units having elapsed since a pre-defined time origin. Then integer arithmetic may be used and, knowing the JDN of the origin, it is easy to calculate the absolute time of any instant.

Access to Current Time

To control temporal behaviour (*e.g.*, to schedule an operation in the future, or to measure time), a control application must also have access to a clock. Such a time source is usually implemented as a counter (also called timer) that counts some basic clock units, realised as relative or absolute time variable. If the time source is only used locally within the control application, the relative notion of time is usually sufficient. In the simplest case, the current time may be represented by a number of basic time units (or ticks) passed from the start-up of the system: Most programming environments include a system function like TICKS that returns this value. A platform may also provide several independent timers with possibly different time units that can be run or stopped.

In addition to measuring the relative time passing, absolute time is often needed to allow temporal synchronisation with events outside of a control system. A typical example is an automatic heating system to be switched on and off at specific hours during a day. In the program, the absolute time is available through system functions like NOW or DATE. A hardware clock is typically used for implementation, which is usually battery-powered to allow for keeping the time even during periods of system shut-down or failure. The clock must only be set once. As an alternative, the absolute time can also be implemented by software. Then an interrupt routine is triggered periodically to increment the seconds, minutes, *etc.* accordingly. Such a clock must be re-set, however, on any system re-start.

For simple control applications, clock accuracies are not so important and small drifts can be tolerated. If higher accuracies are required, time information can be obtained from (or periodically synchronised to) external sources of precise time such as GPS or terrestrial radio stations for time, *e.g.*, DCF 77 [94].

In distributed control systems, additional measures may be taken to assure that the nodal clocks are co-ordinated. This can be achieved in several ways. Global time can be provided from a single system node. This unnecessarily increases the communication load, however, and introduces a potential central point of failure. A better solution is to integrate the clock source and/or clock synchronisation into the communication infrastructure (*e.g.*, synchronisation by reference messages in basic cycles in TTCAN, see Section 3.5.2, Figure 3.9).

Using counters to measure time bears the danger of hidden errors. The first kind of error may occur if counters of too small a size are used. Figure 4.3 shows an example of a delay function. The local variable t is set to a relative time in the future according to the delay desired. Then a loop is performed until this time is reached. The function ticks returns the number of basic time units elapsed since system start. An error may occur, however, if the timer's value is close to its maximum. When the counter reaches its maximum, it resets to zero and starts counting again. If it is accessed twice, prior to and after the turnover, the comparison (and other arithmetic operations) will produce wrong results. In the case of the example given, the loop will be finished prematurely. Furthermore, if the range of variable t is larger than the values provided by the function ticks (*e.g.*, if ticks uses a 16-bit timer and variable t can accept 32-bit values), the situation can emerge that t becomes larger than the function's value and, as the consequence, the loop will never end. Such errors may not occur during application testing. Considering that control systems are often run continuously for long times, however, this may happen later in real use.

```
void delay(int ms)
{
int t;
t = ticks() + ms*TICKS_PER_MS;
while (ticks()<t) { /* void */ };
}
```

Fig. 4.3. Example of a delay function

For illustration, in Table 4.3 the maximum time intervals that can be achieved with different timer sizes are given for the basic time unit 1 ms. Obviously, 32 bits are not enough for continuously running applications. On the other hand, 64 bits are far more than enough, and the implementation of a corresponding temporal arithmetic is rather complex and slow, especially on simple microprocessors. A typical approach to extend the timer size is to use additional variables and software routines. Then, an interrupt routine is triggered when the counter reaches its maximum value and a time variable is incremented accordingly.

Timers with register size of double-words or even greater may give rise to another kind of errors. For example, suppose that a timer can hold 32-bit values. Except on microprocessors with word length of 32-bit or more, reading such a value cannot be performed by a single microprocessor instruction.

Table 4.3. Time delays achieved by counters of different lengths

8 bits:	256 ms $\cong$ 0.25 s
16 bits:	65,536 ms $\cong$ 1 min
32 bits:	$\approx$ 1190 h $\cong$ 50 days
64 bits:	$\approx$ 584,942,400 years

Assume that the 32-bit variable `current_time` is updated periodically within an interrupt routine on every tick of a timer. After the first (lower) part of the variable is read by the main program, an interrupt may occur during which `current_time` is incremented. At a certain instant, the word's lower part may overflow and, thus, be reset, and a one carried to the higher part, *i.e.*, both halves of the timer variable need be updated. After the interrupt, the second half of the timer variable is read. If the initial value of the timer was, *e.g.*, 0x0000ffff, the result of the assignment will be 0x0001ffff instead of 0x00010000. To avoid this problem, reading the timer variable must be placed within a critical section during which interrupts are disabled.

Explicit Declaration of Temporal Requirements

In real-time systems, it should *a priori* be proven that the temporal requirements will be met. The latter are usually defined in off-site documentation. To support automatic schedulability analysis and task scheduling, however, it would be more appropriate to include these specifications in the code. For example, for temporal analysis and for EDF scheduling, the deadlines of a task must be known in advance. An obvious way to achieve this would be to provide in the code an additional attribute for task creation/declaration. If this approach is not applicable due to syntax limitations, strictly formatted information in form of comments ignored by a compiler, but treated by an execution time analyser, can be defined and used for this purpose (see the next section). In addition to task deadlines, some further temporal attributes should be defined like, *e.g.*, minimum time between two occurrences of some signal.

4.1.2 Explicit Representation of Control System Entities

Traditionally, embedded control systems deal with special kinds of objects that are not common to other kinds of applications, *viz.*, tasks, synchronisers, signals, *etc.* Such entities are usually realised inside operating systems and system software, only. To increase the readability of application code and to support a problem-oriented way of thinking, however, it would be helpful if these objects were expressed explicitly within the programming environment.

Representation of Tasks and Tasking Operations

Asynchronous dynamic embedded control systems need to support parallel elaboration of independent activities. To realise the simultaneity requirements, the concepts of tasks, processes, and threads are introduced into application development, representing the basic entities of concurrency. They are usually executed under control of an operating system. In some implementations, however, no operating system is used and the implementation of parallel execution is under control of the application programmers.

Different authors and vendors use different naming conventions for the same concepts. The term process usually represents a functional unit that may invoke tasks or threads. The latter are more or less synonyms. In general, processes are executables with separate memory spaces, while tasks or threads may share common same memory spaces. Here, the general term task is used to denote the unit of concurrency.

Common programming languages make only limited provisions for the declaration of tasks. Usually (like in C and Java), tasking operations and synchronisations are implemented as operating systems calls using one of the predefined tasking models. Examples for the use of system functions to handle tasks (or, in this case, threads) according to the POSIX standard are shown in Figure 4.4

```
pthread_create(thread_handle,thread_attr,start_routine,argument)
pthread_exit (status)
pthread_yield
```

Fig. 4.4. Examples of task handling in POSIX

Routine `pthread_create` creates a new thread and prepares it for execution. Typically, threads are first created in the initialisation phase of program execution. Once created, threads become independent entities and may create other threads during their execution. The first parameter is a thread handle returned by the creation routine and used in subsequent systems calls. The second parameter is a set of thread attributes which can be used to specify the thread's behaviour (*e.g.*, its priority, pre-reserved size of stack, or scheduling policy). The next parameter points to a routine used as a primary execution body of the thread. It is possible that several threads use the same routine, but each thread instance has its own stack. The last parameter is an optional user-defined argument that may be passed to `start_routine`. Routine `pthread_exit` is used to explicitly leave the thread. Typically, it is called after a thread has completed its work and its existence is no longer required. The programmer may optionally specify a termination status. The `pthread_yield` routine forces the calling thread to relinquish processor use, and to wait in the ready queue before it is scheduled again.

However, to support the development of embedded systems better, the entities of concurrency should be declared explicitly (*e.g.*, like tasks in Ada or PEARL). An example of a task declaration in PEARL is shown in Figure 4.5. The portrayed task has priority 10.

```
SomethingToDo: TASK PRIORITY 10;
  do something...
END;
```

Fig. 4.5. Example of a task declaration in the programming language PEARL

If temporal circumstances must be analysed *a priori*, all instances of tasks and their scheduling conditions must be known in advance, and dynamic creation of tasks is not possible. This approach also provides for better understanding of the code, and supports an engineering way of thinking. In a programming environment without explicit task declarations, this can only be achieved if all tasks are created at the start-up of an application.

After a task is instantiated, it can be scheduled for execution. Tasks can be executed periodically or upon occurrences of signals or events. During execution, a task can be suspended until a certain continuation condition is met, or until some shared resource becomes available; an example will be shown in the next section. Again, in simpler programming environments, this is initiated by systems calls for which some appropriate data structures are used. For example, to implement threads periodically executed in a pure POSIX routine, the programmer must either allocate a system timer and bind it to thread execution, or implement an infinite loop with an appropriate delay within any cycle. PEARL is one of the few programming languages providing explicit tasking operations. Some examples for tasking operations and scheduling conditions are shown in Figure 4.6.

In the first example, the task is activated periodically every hour until a certain absolute time. In the second, the task is activated asynchronously by an external event, *viz.*, when a push-button is pressed, after a delay of 100 ms. The third example starts a periodic task activation beginning at 6:30 hours every 2 min for a time interval of 15 min. The example with the RESUME statement suspends the execution of the current task for 5s. The next two statements suspend the mentioned task, and schedule its continuation after 10 min, respectively. A TERMINATE statement prematurely ends the execution of specified tasks. A PREVENT statement annihilates the schedules for all tasking operations with respect to a specified task. For example, by this a periodic task will not be activated anymore. In contrast to task termination, however, PREVENT does not affect the execution of tasks already started.

```
EVERY 1 HRS UNTIL 12:00 ACTIVATE SomethingToDo;

WHEN ButtonPressed AFTER 0.1 SEC ACTIVATE ButtonIsPressed;

AT 6:30 EVERY 2 MIN DURING 15 MIN ACTIVATE Wake_me_up;

AFTER 5 SEC RESUME;

SUSPEND SomethingToDo;

AFTER 10 MIN CONTINUE SomethingToDo;

TERMINATE SomethingToDo;

PREVENT SomethingToDo;
```

Fig. 4.6. Some tasking operations in the programming language PEARL

4.1.3 Explicit Representation of Other Control System Entities

In addition to tasks, other items typical for embedded systems should be dealt with explicitly by the programming environment. Examples for them are signals and synchronisation mechanisms. In reality, availability and use greatly depend on capabilities of the underlying operating system. Different operating systems provide different kinds of synchronisers. For example, POSIX knows three kinds of synchronisers: semaphores, mutexes, and condition variables, each with its own set of system routines. A set of semaphore functions is shown in Figure 4.7 for illustration,.

```
int sem_init(sem_t *sem, int pshared, unsigned int value);
int sem_wait(sem_t *sem);        /* P(sem), wait(sem) */
int sem_post(sem_t *sem);        /* V(sem), signal(sem) */
```

Fig. 4.7. Basic system functions for semaphore operations in POSIX

An example for the opposite approach is the programming language PEARL which allows one to declare explicitly semaphores and bolts (synchronisers with both exclusive and sharing locking capabilities) (see Figure 4.8) and treat them within the language.

To prevent difficulties that may occur with task synchronisation, corresponding mechanisms should be structured, *i.e.*, it must be clear where a section for mutually exclusive use starts and where it ends. A scenario, where a synchroniser is locked in one task and then unlocked in another, produces complex behavioural patterns and may lead to deadlocks. The same holds

```
DECLARE sem SEMA PRESET(1);   ! Declare and preset a semaphore
...
REQUEST sem;      ! P(sem)
...
RELEASE sem;      ! V(sem)
```

Fig. 4.8. Basic semaphore operations in PEARL

if within a single task a synchroniser is locked in one control structure and unlocked in another.

```
if (SomeCondition)
   wait(Semaphore);


...


if (SomeOtherCondition)
   signal(Semaphore);
```

Fig. 4.9. Example of ambiguous use of synchronisers

In Figure 4.9 a scenario with unsafe handling of synchronisation mechanisms is illustrated. According to the conditions of if-statements, two erroneous scenarios are possible: the synchroniser is locked (wait successful), but never unlocked (missing signal), and the synchroniser is only unlocked (no wait). With structured synchronisation constructs such as monitors (see Section 2.3.4) it is possible to prevent deadlocks, because the sequences of lock and unlock operations are always in order and deterministic. In the case of synchronisation mechanisms with unstructured syntax, it is usually left to the programmer to assure this in the code.

4.1.4 Support for Temporal Predictability

The ultimate goal in programming hard real-time control systems is to assure that all activities will be terminated within predefined periods of time and to assure this *a priori*. In this sense, all parts of control applications — including program code — must behave deterministically in time. Consequently, execution times of all programming constructs must be deterministic and easily predictable. For example, synchronisation mechanisms usually include statements for locking with indefinite durations. In real-time systems, such statements must be temporally guarded not to take more than allocated amounts of time. Similarly, there can be no statement in a temporally critical tasks that waits for a user intervention an indefinite long period of time.

The temporal behaviour of some other programming elements, like dynamic data structures or recursion, is difficult to predict. This is because their actual complexity depends on current data values and cannot realistically be determined in advance. With reasonable effort, dynamic data structures can be transformed into static ones, and recursive algorithms can be implemented in an iterative manner. For the same reason, databases, expert systems, and similar programming tools should be avoided, unless they are specifically designed for use in hard real-time tasks. Further, some other features, such as unrestricted use of GOTO statements, make proper temporal analysis impossible.

Assurance of temporal predictability must be integrated into the development tools. For example, a compiler should notify programmers when temporally unsafe features are used. Temporal analysis can generally not be carried out by hand; it requires proper software tools performing the analysis automatically. More details about this topic will be given in the next section.

4.1.5 Support for Low-level Interaction with Special-purpose Hardware Devices

In typical desktop applications, the diversity of hardware components is limited and the programming environments provide high-level access to them. In contrast, in embedded systems there is a tight correlation between the hardware and software parts of control applications. A vast number of different hardware components is used, which are not generally recognised by the system software. Thus, only low-level interaction is possible. In general, all hardware devices are directly accessible through some sorts of input/output interfaces, through which control applications interact with hardware devices (*e.g.*, a temperature sensor's analogue output is directly attached to an analogue-to-digital converter, whose register is, in turn, accessible by the processor).

Input/output interfaces expose themselves to programmers through sets of control, status, and data registers that are (at least in simpler cases) mapped into the memory spaces of processors. To be able to perform adequate data handling, programming environments must support direct access to these registers (*i.e.*, direct access to some memory locations). Not all programming languages allow this kind of interaction. In Modula-2, however, a variable can be associated with an absolute address using the `MAKEADR` operator:

```
VAR AD [MAKEADR(0B800H)] : Integer;
```

Some programming languages provide limited capabilities to extend the handling of hardware registers by automatically generating low-level device drivers. For example, in PEARL it is possible to declare a hardware register as an input/output device, and use similar input/output statements as with any other device supported by the operating system.

As another example, in ANSI C it is not possible to associate a variable to an absolute memory location, but to set a value of a pointer to reference a specific memory location. Status and control registers usually consist

of separate bits or groups thereof, that can be accessed individually. Therefore, it is beneficial for programmers if such fields can be declared explicitly.
In Figure 4.10, an example of low-level access to an A/D converter in the
programming language C is demonstrated. Here, ADCON is a control/status
register consisting of several bit fields described by the data type adcon_t.
The bit fields are used to allow access to individual sections of the register.
The union is used, however, to access the register as a whole. The register is
mapped to memory location 1234 (hexadecimal). For convenience, the macro
ADCON is defined that simplifies programming and improves readability of
the code by inherently resolving all typecasting.

```
typedef union {
  struct {
    ADON unsigned int:1;        /* 1: enable A/D converter */
    unsigned int:1;             /* unused */
    GODONE unsigned int: 1;     /* (write)1: start conversion,
                                   (read) 1: converting, 0: finished */
    CHS unsigned int:3;         /* input channel */
    ADCS unsigned int:2;        /* clock source for the conversion */
  };

  content unsignedchar;

} adcon_t;

#define ADCON *((adcon_t *) 0x1234)
  ...
/* initialisation */
ADCON.content= 0xC1;

/* start of A/D conversion */
ADCON.GODONE= 1;

/* waiting for A/D conversion to be finished */
while(ADCON.GODONE==1)
    { /* wait */ };
```

Fig. 4.10. Example of direct access to register of an A/D converter in ANSI C

Caution must be taken in accessing the registers of input/output interfaces.
In contrast to basic program variables, their contents may change independently of program code execution. In the previous example, the GODONE bit of
the ADCON register is reset to zero when A/D conversion finishes. Difficulties
arise if optimisation is used during code compilation. Analysing the code, the
compiler may assume that the GODONE bit is set to one permanently. Because
of that, the condition in the subsequent while loop would always be true, and

the loop would be executed for ever. Consequently, the loop's condition testing may be eliminated from the generated code by an optimising compiler. To avoid this, the variables that may change their values asynchronously must be specially marked. In most programming languages, such variables are marked with the "volatile" attribute.

Another scenario where program code interacts with hardware on a low level are interrupt service routines (ISR). Interrupts are a part of a mechanism provided by the majority of processor architectures to allow efficient supervision of input/output devices. Interrupt signals are generated by peripheral interfaces (or other hardware devices, *e.g.*, timers). They signal to processors that some events occurred, which need attention. When an interrupt occurs, the processor temporally stops to execute the current application, saves its context, and starts an interrupt service routine. Each interrupt source may have a separate interrupt service routine. Based on the discussion in Section 3.2, for easier temporal predictability and better fault tolerance, it is wise to prevent interrupts altogether and to use polling techniques instead. Interrupt service routines also strongly depend on the hardware devices used. Sometimes, however, it is not necessary to write them, *e.g.*, when very short reaction times upon external events are required, or when special-purpose hardware is used to which software support has no direct access.

Interrupt service routines have a similar structure as procedures, but without parameters. When an ISR terminates, however, a different return mechanism than for normal procedures must be used, because the processor must restore the register context saved at the start. Therefore, an additional attribute is needed to indicate this to the compiler. In the language C, an ISR is usually declared as:

```
interrupt isr() {
    ...
}
```

In practice, however, the syntax of ISRs is an extension to the standard and not all compilers support it. Another mechanism to implement ISRs is to write a standard procedure and to call an appropriate system function, that creates a wrapper around the code and provides for all particularities of interrupt handling. In POSIX, *e.g.*, this is implemented with the system routine posix_intr_associate(). Another possibility is to write an ISR in assembly code and to link it with the main application. If neither possibility applies, an appropriate device driver must be implemented or provided to be integrated into the operating system.

Interrupt service routines run in privileged modes. Any error that occurs in them may have catastrophic consequences to system operation. Therefore, ISRs must be well verified and tested. Moreover, they should be as short as possible to minimise the delays caused to the applications interrupted. If longer servicing is required, it can be split in two parts. The first part is a short IST scheduling the second part in form of a task, which carries out the actual servicing. The task is executed under the normal tasking mechanism.

Another approach to low-level interaction with the hardware is by inserting portions of assembly code. Most programming environments for embedded systems allow one to insert parts of code written in the native assembly language of the respective microprocessor. Again, this is usually an extension to the syntax of the language used, and is handled differently in different tools. By this, programmers can interact with the hardware directly, and have full control over execution. However, the code inserted is implementation-dependent and difficult to maintain.

4.1.6 Support for Overload Prevention

Theoretically, if formal schedulability analysis can be performed, deadlines of tasks should always be met. In real situations, however, not all circumstances of system execution can be predicted with sufficient confidence. In addition, hardware failures or transient increases in the occurrence frequencies of input signals can result in deadline violations. With some additional support provided by the applications, deadline misses can be prevented.

Any time a new task becomes ready, the operating system re-schedules all tasks in the system and analyses schedulability at run-time. Using a proper scheduling policy (*e.g.*, earliest deadline first), the operating system can thus predict possible timing errors that are going to happen in the future. If such an overload situation is detected, newly arrived tasks with longer execution times can be replaced by shorter ones to satisfy all deadline conditions.

As deadline prevention is under control of the operating system, all possible execution alternatives for tasks must be known to the latter in advance. To this end, all alternative implementations of a task must be declared. Unfortunately, no commercially available development environment supports this approach. A prototype solution will be demonstrated in Section 4.3.2.

If a deadline is violated because a sporadic event is firing more frequently than anticipated, two other measures can be applied:

- If the event originates from a hardware interrupt, the interrupt can be prevented from occurring by manipulating the associated device control registers, or
- Interrupts can be prevented altogether and replaced by periodic polling of data.

4.1.7 Support for Handling Faults and Exceptions

To provide for general fault tolerance, the techniques described in Section 1.3.3 should be applied. This kind of fault management is usually employed separately from direct application programming, because failures of processing nodes, plant components, or other hardware components must be handled globally. There are certain other situations, however, where faults can (and should) be handled by the applications. Such situations are failures within certain subsystems, exclusively controlled by application programs, which, if handled locally, must not always be reported to the outside, thus relieving the global systems from servicing them.

Typical examples for this are faults induced by software errors. If not handled properly, these faults may result in chaotic behaviour of the applications. In the case of desktop applications, their negative consequences are, in general, relatively harmless. On the other hand, for safety-critical control applications, such faults could result in unacceptable losses.

Apart of software errors, applications may also deal with faults on locally attached hardware devices (*e.g.*, a fault of a sensor directly attached to a processing node) either by handling them directly or by notifying higher-level fault-management systems.

In real-time systems, a delay in the delivery of results must also be considered as fault. As such, it may as well be handled within applications.

Handling of faults with appropriate service software constitutes one of the most severe threats to temporal determinism. Thus, dealing with exceptions has to be elaborated very thoughtfully in real-time environments.

An *exception* is an exceptional event, an action, or a condition in the execution of a program that changes its normal flow. A closer examination of the possible sources of exceptions in embedded systems leads to the following classification into three categories: *preventable, avoidable, and catastrophic exceptions.*

Preventable Exceptions

The best way to handle exceptions is to prevent them from occurring in the first place — as far as this is possible. Exceptions are preventable by imposing certain restrictions on the use of potentially dangerous system features. Some features, which could trigger exceptions as consequences of irregular situation are presented below.

Memory-related exceptions. When using reference pointers and dynamic data structures, address error exceptions may be triggered in numerous situations, *e.g.*, by referencing non-existing or system-protected parts of memory (because of applying a not initialised or NULL reference). Further, when a new portion of memory needs to be allocated for some dynamic data structure, it is possible that no free memory space is available. Similarly, unlimited use of recursion can cause stack overflow. And even if no errors occur, the temporal behaviour of these features is difficult to predict. Hence, languages for programming high-integrity applications usually prohibit these features (see, *e.g.*, MISRA-C [84]). If not imposed by the compiler, memory-related exceptions can be prevented by proper programming and discipline. All memory needed for the execution of an application should be allocated during initialisation. The use of stacks must be analysed and enough memory space allocated for them to accommodate both parameter passing and memory allocation for local variables.

In modern microprocessor systems virtual addressing provides for virtually unlimited memory space for any application. The secondary (mass) storage devices are used to handle pieces of data or program memory that momentarily cannot reside in main memory. Unbounded delays are introduced when swapping between main and mass storage occurs. Considering the current capacities of memory chips, there is no more need for virtual addressing in embedded systems. By renouncing it, related faults and temporal uncertainty are prevented.

Peripheral-device-access-related exceptions. Invalid device addressing and access conflicts can be prevented by only using statically mapped I/O devices known at compile-time, which must be declared and should be checked by the compiler. A good example for this is the specification of configurations in the so-called "system parts" of PEARL programs where all hardware devices used in the algorithmic parts of applications must be defined. It allows for system configurations to be verified at initialisation time. In case of device loss or auto-test failure, start-up can be aborted.

Data-related exceptions. The exceptions of this group are caused by data illegal for some operations (*e.g.*, square root of a negative argument). Similarly, range overflow or underflow in arithmetic operations may occur when the result of an operation is too large or too small, or when high-precision data is mapped into small-precision variables. Yet another source of data-related exceptions are situations when array indices exceed their bounds because of declaration/run-time value mismatches.

To prevent undesirable events of the latter group, various measures can be taken. The obvious possibility is to implement strict type checking in the language used, so that possible irregular operations can be detected and reported already at compile-time. An example for this approach is the exception-free language NewSpeak [25] supported by an appropriate safe architecture. The main characteristic of NewSpeak is strict definition of data types. In an assignment statement, for example, the data type of the variable on the left-hand side must include all possible values that may be produced by the expression on the right. However, since strict type checking may become very complex and, thus, often impractical, the principle of extending the IEEE standard for binary floating-point arithmetic [15] should be employed: the input and output data types are extended by two "irregular" values representing "signed infinity" to accommodate overflows and underflows, and "undefined (not-a-number — NaN)" to formalise the results of invalid operations. The latter is used when a non-recoverable problem occurs in a calculation. Thus, the generated irregular results do not raise exceptions, but are propagated to subsequent or to higher-level blocks, which may be able to handle them.

As the last links in propagation chains of such irregular values, intelligent I/O interfaces should react to them in predefined ways. For instance, instead of an "infinite" analogue value, the maximum possible voltage may be generated by an intelligent D/A converter as result of a certain control algorithm. Precautions should be taken, however, if such a value persists for a longer time, indicating that the controlled system is not responding, and that the infinite value is likely to be a consequence of a computing error.

Avoidable Exceptions

There are situations where unusual (or irregular) events cannot be prevented from occurring, but it is known in advance that they might occur, and it is possible to detect their occurrence. People tend to deal with unwanted events as exceptions. It is, however, irrelevant whether they are unwanted or not. If they can be anticipated during the design phase, they should be included into the specifications to be handled adequately.

Thus, dealing with them becomes part of the application software. Common examples of this kind of events are tasking errors, *e.g.*, activating a non-existent task. Such exceptions can be avoided by prophylactic run-time checks before entering critical operations. Many tasking errors are also avoidable by using monadic operations to interrogate current system states.

Errors in I/O devices also fall into this class. Intelligent, fault-tolerant and self-checking peripheral devices shall be able to recognise own malfunction. They shall have as much intelligence as possible to reasonably react in situations of conflict. In case of recoverable errors, they should try to recover locally. If stand-by redundancy is available, it can be switched over to redundant units. The temporal behaviour of these actions, however, must be deterministic. For calculations of execution times, worst-case values have to be taken into account.

Coping with Catastrophic Exceptions

Unfortunately, in industrial embedded systems situations often occur which can be neither anticipated nor avoided. These are situations when "the impossible happens" [9], when programs do not follow their specifications due to hardware failures, residual software errors, or wrong specifications. For example, failure of a part of data memory could cause a change of a value, which was declared to be a constant. It is also possible that the behaviour of the environment was not estimated realistically, and is requesting more frequent service than originally specified resulting in system overloads.

According to a reference study in the domain of non-preventable exceptions [23], exceptional situations can be dealt with by programmed exception handling, and by default exception handling based on automatic backward or forward recovery using recovery blocks. Since programmed exception handling should be included in the requirements of embedded hard real-time systems, it can, thus, be treated as normal functionality.

The principle of backward recovery is to return to a previous, consistent system state after an inconsistency is detected by consistency tests called post-conditions. This can be performed in two ways, *viz.*,

- By the operating system recording the current context before a program is run, and restoring it after unsuccessful termination, or
- By recovery blocks inside the context of tasks whose syntax reads as shown in Figure 4.11, where P_0, P_1, *etc.* are alternatives which are tried consecutively until either consistency is ensured by meeting the `post-condition`, or the segment `failure` is executed. Each alternative should independently be capable to ensure consistent results.

```
ensure post by P₀

else by P₁

else by . . .

else failure
```

Fig. 4.11. Principle of recovery blocks

For embedded systems, a severe problem arises if there are any actions triggered, like commencing a peripheral process, which cause an irreversible change of a failed alternative's initial state, rendering backward recovery generally impossible. Therefore, no physical control outputs should be generated inside the alternatives which may require backward recovery in case of failure. Then, only forward recovery is possible, bringing the system to a certain predefined, safe and stable state.

The alternatives of backward recovery should contain diversely designed and coded programs to cope with specification errors, and to eliminate possible implementation problems or residual software errors. They may employ alternative design solutions or redundant hardware resources, when problems are expected. A further possibility is to assert gradually less restrictive pre- and/or post-conditions and, thus, to degrade performance gracefully in the case of exceptional situations.

The technique of forward error recovery tries to obtain a consistent state from partly inconsistent data. Which data are usable can be determined by consistency tests, error assumptions, or with the help of independent external sources.

Lower-level Exceptions

Exception handling is a mechanism to deal with exceptions in the programming language both syntactically and semantically. Exceptions that may occur should be anticipated in specific parts of code, and dealt with by using appropriate handlers, *i.e.*, parts of code that are executed only when exceptions occur. Simpler programming languages do not provide for exception handling at all, or include it in a non-structured way like if any kind of exceptions occurs anywhere in a program block that includes an **on error** statement, execution is transferred to a labeled statement:

 on error goto exception-handler-label

Modern programming languages implement structural exception handling where code of one exception handler can be nested inside that of another one. If an exception occurs, it is first handled by the innermost exception handler. If it cannot be handled at that level, the next outer exception handler is called. Only at the end, if no exception handler can deal with a specific exception, it is dealt with by the operating system.

The sources of exceptions can be classified into different categories, *e.g.*, arithmetic or I/O exceptions. These classes can further be refined into more and more detailed ones until specific exceptions, *e.g.*, division by zero in the class of arithmetic exceptions, can be expressed. In this way, programmers can deal with different classes of exceptions individually, both generally for entire exception categories and specifically for individual exceptions. Usually, programmers may define new classes of exceptions and invoke the exceptions by software.

```
try
    program-code

catch (exception-class-1)
  exception handler for exception-class-1

catch (exception-class-2)
  exception handler for exception-class-2

finally
  final block statements

end
```

Fig. 4.12. General form of structured exception handling

The most general form of exception handling is shown in Figure 4.12. If an exception occurs in the program code following the **try** clause, which is being protected, one of the exception handlers is executed. Based on the exception's kind, the first exception handler matching the exception's class is called. Exception handlers should be arranged from the most specific to the most general one. The statements of the **finally** block are executed regardless of whether an exception has occurred or not. This can be used to re-arrange some resources allocated within the basic programming block, which may not be needed anymore, because the program block terminated or an exception occurred.

4.1.8 Support for Hardware/Software Co-implementation

Not all parts of control applications are necessarily implemented in software. For reasons of decreasing response times or of improving temporal predictability, verifiability, and fault tolerance it is common to use hardware solutions like System-on-Chip (SoC) and Field Programmable Gate Arrays (FPGA). However, no commercially available development tools currently allow for large-scale complex hardware/software co-design and implementation. The decision as to which parts of applications are to be implemented in software and which ones in hardware is left to designers and programmers. Then each part is developed separately. As application code functions of operating systems and other system software can be implemented in hardware as in the hardware architecture described above.

There are several hardware description languages (HDL), primarily used to describe electronic circuits formally. With such languages a circuit's operation, design and organisation are described. Tools based on them perform tests to verify operation by means of simulation. The semantics of the languages also include explicit notations to express timing and concurrency.

An example of a programming language that has been devised to describe both the software and the hardware parts of systems is SystemC [20]. It provides hardware-oriented constructs within the context of C++ as a class library implemented in standard C++. Programs written in SystemC can be translated either to standard C (for the software components) or to a hardware description language (for the hardware components).

An example of how a programming language can be extended to define various parameters of distributed control applications to be run on an architecture similar to the one of Section 3.3.1 was given in [19]. The specification language is based on PEARL for Distributed Systems [30] and is, actually, more a specification than a programming language. Indispensable to specify the behaviour of distributed systems, it contains language constructs to describe

- Hardware configurations,
- Software configurations,
- Communication configurations and their characteristics (peripheral and process connections, physical and logical connections, transmission protocols), as well as
- Both conditions and methods of carrying out dynamic reconfigurations in cases of failure.

In contrast, the language contains only a few executable constructs in the classical sense, *viz.*, just to exchange messages. The example in Figure 4.13 demonstrates how a hardware configuration can be specified using the proposed notation.

Processing nodes are defined in the **STATIONS** section. For each processing node, its type, amount of physical memory available, devices connected,

```
ARCHITECTURE;

STATIONS;
  NAME: TP0;
  PROCTYPE: M68307 AT 16 MHZ;
  WORKSTORE:
    SIZE 128*1024;
    SPACE '00000'B4-'1FFFF'B4 READONLY WAITCYCLES 6;
    SPACE '20000'B4-'3FFFF'B4 READWRITE WAITCYCLES 2;
  DEVICE I2CTP0: PERIPHERAL;
  STATEID: ( NORMAL, CRITICAL );
STATEND;

CONFIGURATION;
 COLLECTION TP0NORMAL
   MODULES (
     TP01         IMPORTS  (F001, F002) FROM LIBRARY
                  EXPORTS  (INITTP0),
     TP02         IMPORTS  (F001, F002) FROM LIBRARY
                  EXPORTS  (CALCULATE),
     LIBRARY      EXPORTS  (F001, F002));
   TASKS (
     INITTP0    ON START DEADLINE 2000,
     MAINTP0 EVERY 0.1 SEC DEADLINE 1000);
   PORTS (TPLINK0);
   LOAD NORMALTP0 TO TP0;
   CONNECT TPLINK0<->KPLINK0;
 COLLECTION TP0CRITICAL
   ...
 COLLECTION TP1NORMAL
   ...
CONFEND;

NET;
  TPLINK0<->KPLINK0;
  TPLINK1<->KPLINK1;
  TPLINK2<->KPLINK2;
NETEND;

KERNELPROCESSOR;
  NAME: KP0;
  SCHEDULING EDF;
  TICK 0.01 SEC;
  MAXTASKS 50;
  MAXSEMA 32;
  MAXEVENT 31;
  MAXEVENTQUEUE 256;
  MAXSCHEDULES 64;
KERNELPEND;

PERIPHERALS;
  NAME: AD1;
  INTERFACES (
    ADSTART OUT BIT(1) I2C ADDR 102,
    ADDATA  IN FIXED(7) I2C ADDR 103);
  CONNECT AD1<->I2CTP0;

PERIPHEREND;
```

Fig. 4.13. Example system and hardware specification expressed in a high-level programming language

and possible operation modes are described. The attributes needed for execution time analysis are also defined. In the CONFIGURATION section, all possible software collections are defined. Collections are the basic units of applications that are loaded to different processing elements during execution. Each collection consists of a set of software modules and their interfaces. Further, a set of tasks with scheduling policies and deadlines is defined. Finally, intermodule connection links and a loading scenario are declared. Physical interconnections between processing elements are described in the NET section. The KERNELPROCESSOR section defines the properties of operating system and system software (here, an architecture with central kernel processor is explained). In the PERIPHERALS section, various input/output peripheral interface (together with their physical characteristics) used by the application are defined.

4.1.9 Other Capabilities

There are other properties desired of programming environments that do not only apply to embedded control system applications. Programming environments should, for instance, be user- and problem-oriented, should support re-usability and the development of large-scale applications, or should have efficient debugging capabilities.

4.2 Time Modeling and Analysis

Time is obviously one of the most important aspects of real-time systems, where it manifests itself in several ways. On one hand, time is an integral part of system specifications. There are response deadlines and maximum frequencies of event occurrence. On the other hand, there are temporal system properties which are needed to satisfy these specifications.

For real-time systems it is imperative that overall system capabilities allow one to fulfil all predefined requirements. In multiprogramming environments, the main requirement can be expressed as *schedulability, i.e.*, the ability to find a schedule such that each task will meet its deadline [108]. A necessary pre-condition for this is *temporal predictability of execution times, i.e.*, the capability to estimate correctly their worst-case values. In theory, it is possible to observe various other aspects of time, *e.g.*, minimum or average execution time. For obvious reasons, however, only maximum or worst-case execution time (WCET) is relevant for hard real-time systems.

There is a host of contributions to the field of WCET analysis, *e.g.*, [66, 69, 95, 86, 89, 88], and later [46, 7]. Most of them are based on supplying some additional knowledge about programs to analysers in order to achieve better estimation of run-times. Although yielding good results, which are close to the actual values, these solutions do not appear to be sufficiently practical, yet, as too many factors influence the overall execution times of any application:

Temporal determinism and predictability of application code. To enable execution time analysis, no time-critical code section may take indefinitely long to execute (*e.g.*, infinite loops, or waiting for operator input). Furthermore, to allow for realistic code analysis, some programming constructs are undesirable. For example, unbounded recursion, unbounded loops, dynamic data structures, *etc.* are very difficult, if not impossible, to evaluate properly.

Capability of processing elements to execute application code. In general, faster processors can execute tasks in shorter time. More sophisticated processors making use of caches, massive parallel execution of instructions, *etc.*, also yield shorter average execution times. The very same architectural features, however, also render the estimation of worst-case execution times very difficult.

Capabilities provided by operating systems. An operating system on which an application is executed always imposes some limitations: minimum quantum of time, precision of activation instants of tasks, granularity of timers and their maximum intervals, *etc.* Tasks scheduling and synchronisation operations also cause unpredictability in timing, and introduce additional increases in execution times. Other operating system services, like memory management or file system handling, further contribute to temporal uncertainty, which is very difficult to evaluate.

Capabilities provided by communication infrastructure. In distributed systems, the communication channels always introduce some kind of delays, and may produce jitter in system behaviour. Furthermore, the typical and most commonly used communication protocols do not guarantee upper bounds for the delivery times of data, which makes them unsuitable for hard real-time implementations.

In conclusion, worst-case execution times can properly be estimated only if all components of control systems (*i.e.*, hardware, system software, and application code) behave temporally predictable. With modern microcomputer architectures, this turns out to be very complicated. Too pessimistic evaluations could characterise systems as unfeasible, and require unreasonable and unnecessary effort to improve their performance. On the other hand, too optimistic estimations could cause deadline violations when applications are executed in real systems.

There are several stages in the development of control systems when timing analyses should be performed. First, the timing properties of specifications are evaluated, then those of hardware architecture and application source code. When hardware and operating system are available, the executable product from the source code can be evaluated. Each step produces more and more realistic results. It is important to detect any potential problems as soon as possible in the design. The earlier inconsistencies are detected, the easier and cheaper they can be corrected.

The basic idea behind the WCET analysis of a task is, first, to evaluate or estimate the maximum execution time of the atomic elements contained. Sequences of such elements form basic blocks. Then, knowing the structure of the task, it is possible to derive recursively the execution time of its entire code by combining the basic blocks in hierarchical compound blocks. There are three basic ways basic blocks can be combined into compound blocks: sequence, alternative, and iteration.

In the first example (Figure 4.14), the compound block (CB) consists of two basic blocks (B1 and B2). First, the block B1 and then the block B2 is executed. The arrows represent entry and exit points of the blocks. The WCET analysis starts by evaluating the execution times of the atomic blocks (*e.g.*, adding the execution times of subsequent instructions). Then, the total WCET of the compound block is the combination (sum) of the WCETs of B1 and B2:

$$t_m(CB) = t_m(B1) + t_m(B2)$$

The expression t_m in this and the following equations represents the maximum (worst-case) execution time.

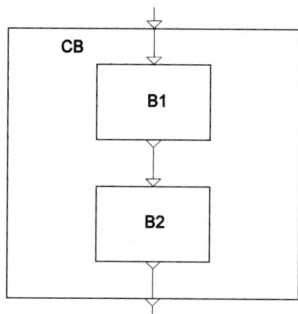

Fig. 4.14. Example of the sequential combination of basic blocks

In the second example (Figure 4.15), the compound block also consists of two atomic blocks. Here, however, either B1 or B2 is executed, but not both. After the WCET analysis of the atomic blocks, the total WCET of the compound block equals the greater one of the worst-case execution times of blocks B1 and B2:

$$t_m(CB) = max\{t_m(B1), t_m(B2)\}$$

The same is true, if alternatively, both B1 and B1 are executed in parallel.

The next example (Figure 4.16) shows a loop-back data flow. The inner block (BL) is the loop body, which is executed several times before the CB finishes. Here, it is necessary to know the maximum possible number of the inner block's repetitions. It does not matter if sometimes (or even in most

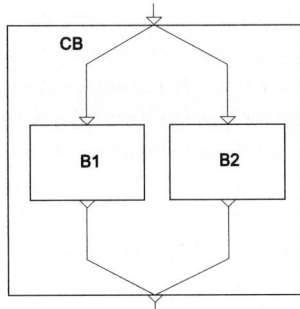

Fig. 4.15. Example of a compound block with alternatives

cases) the loop is finished earlier, since the worst case is of interest, only. Then the WCET of the compound block is the product of that of the inner block and the maximum number of iterations:

$$t_m(CB) = t_m(BL) \times i_m(BL)$$

Here, i_m denotes the maximum number of block iterations.

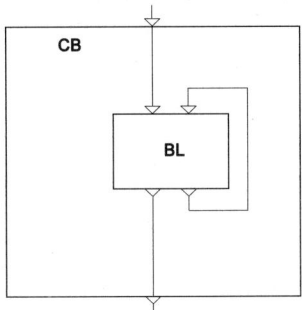

Fig. 4.16. Example of a compound block with a loop

4.2.1 Execution Time Analysis of Specifications

The first estimation of a control system's timing conditions can be derived from its specifications. Since the implementation details of the final system are not known at that stage, the actual task execution times cannot be determined, yet. At best, experience from similar, already finished projects can be used. Nevertheless, based on the tasks' activation conditions and deadlines, it can be assessed if the specifications are plausible, and whether there are potential bottlenecks that could result in deadline misses. It would be very helpful if some kind of formal notation were used for specifications. If a formalised pseudo-code is used to assert temporal information, for instance, techniques for source code examination (described in the next section) can be employed.

In control engineering praxis, formalised functional models are frequently used to portray control systems. These models consist of sets of function blocks, each taking several inputs and producing outputs. The blocks' functionalities are usually defined by means of mathematical functions. Owing to their simplicity, it is relatively easy to estimate the execution times of individual function blocks, and then derive the execution times of tasks constructed out of them. Later in development, these function blocks are directly translated into appropriate code segments, and techniques of analysing the execution time of program code can be used to improve the earlier estimations. Near the end of the design phase, when the hardware platform is constructed, the code blocks' execution times can directly be measured, and the timing analyses further refined.

4.2.2 Execution Time Analysis of Source Code

Execution times can be more precisely estimated when application source code is evaluated. For this a kind of pre-processor is needed that takes an application's original source files as inputs, extracts the information relevant for timing analysis, and employs a method of WCET analysis. To enable this, some limitations must be imposed on the program structure. These limitations will be explored in Section 4.2.6. Here, some basic strategies for WCET analysis of source code are described. The same techniques can be used with any kind of application description that uses a program notation (*e.g.*, with machine code instructions).

Expressions

Expressions are atomic elements of source code. Some examples for expressions are variable accesses, constants, mathematical and logical terms. Estimating their execution times is straightforward if the execution times of the operators are known for specific processors. However, several factors cause difficulties in the evaluation. First, an expression can be modified by a compiler performing different optimisation techniques. Multiplication by a factor of two, for example, can be replaced by a much faster addition:

$$a \times 2 \Rightarrow a + a$$

On simpler processors, multiplication takes much more time than addition. If a microprocessor does not implement multiplication, it is carried out by software routines. Second, modern microprocessor architectures utilise parallel execution of instructions to a high degree. In most cases, this yields much shorter execution times. Further, some operators are data-dependent, *i.e.*, the execution times depend on the operands' current values, which are not known at the time of analysis.

All of these (and other) causes usually have relatively little influence on the accuracy of the execution time estimate for a specific expression. If the expression is a part of frequently executed code (*e.g.*, inside a loop), however, the accumulated error can be significant. Therefore, if realistic execution time analysis from source code is a goal, simpler processor architectures need to be employed.

Assignment Statement

One of the simplest statements in any computer language is the assignment:

$$s_a ::= \texttt{l--value '=' r--value}$$

Execution time analysis considers here the time to evaluate the expression on the right side, the time needed to calculate the target address of the left side (*e.g.*, if it is a structure or array reference), and the time needed to transfer the calculated value to the target.

Function or Procedure Calls

A procedure call requires evaluation of the parameters, transfer of the latter to the procedure, and invoking the procedure's code:

$$s_p ::= \texttt{call proc } (p_1, \ p_2, \ \ldots, \ p_n)$$

Therefore, the corresponding execution time can be calculated by adding the execution times of the mentioned activities. In addition, care must be taken for different kinds of parameter transfer (*i.e.*, transfer by value or transfer by reference), the mechanism of parameter transfer into the procedure (*i.e.*, by stack, by register, or a combination of both), the housekeeping of memory implied by the call (*i.e.*, by caller or by callee), *etc.* In case of a function call, the function's result is part of an expression, and the call's execution time is added to the one of the expression.

Another situation arises when system routines are called (*e.g.*, operating system routines). Then, execution time analysers have no access to the source code. Therefore, the execution times of such routines must be determined separately. Besides that, some system calls may lead to diverse execution times.

Sequence of Statements

$$s_s ::= s_1 \ ; \ s_2 \ ; \ \ldots \ ; \ s_n$$

The most fundamental compound structure in programming languages is the sequence of statements. It represents a linear flow of code. Since the statements are executed one after the other, the total WCET of a sequence is the sum of each individual statement's worst-case execution time:

$$t_m(s_s) = \sum_{i=1..n} (t_m(s_i))$$

Even in this simple case, however, some difficulties can arise if there is a dependency between the execution times of subsequent statements (*e.g.*, shorter execution time of one statement may yield longer execution time of the subsequent one, and *vice versa*). In automatic WCET analysis, the worst estimations of both statements are taken. This usually happens due to data dependency between the statements. Practical example of this scenario will be given later.

Alternative Statements

All programming languages allow some sort of decision points where one of several alternative routes through a code is selected based on some condition. The most basic form of alternatives is the if-statement:

s_{if} ::= if condition_expression then s_{then} else s_{else}

First, the condition is evaluated. Then, based on its value, one of the two alternatives is executed. At first glance, the execution time equals the execution time of the longer alternative. For accurate estimation, however, the time needed to evaluate the condition must also be taken into account. This execution time is often much greater than that of the statement block following, as it may, for example, include a call to a complex function:

$$t_m(s_{if}) = t_m(condition\_expression) + max\{t_m(s_{then}), t_m(s_{else})\}$$

Other forms of alternatives (*e.g.*, the switch statement) are evaluated in a similar way, except that more alternatives are to be considered:

```
s_case  ::= expression of
             alternative_1 : s_1 ;
             alternative_2 : s_2 ;
             ...
             alternative_n : s_n ;
          end case
```

$$t_m(s_{case}) = t_m(expression) + max_{i=1..n}\{t_m(s_i)\}$$

For precise analysis, the hidden administration time due to the jumps between the alternatives must also be considered.

Iteration

In general, the maximum execution time of a loop is equal to the execution time of the loop body multiplied by the maximum number of loop iterations. Therefore, the basic prerequisite to evaluate a loop's WCET is to know the maximum number of iterations in advance. For the for-loop statement with constant bounds:

s_{for} ::= for id = lower_bound to upper_bound do
> s_{body}
> end for

the WCET can easily be calculated as

$$t_m(s_{for}) = t_m(s_{body}) * max\{upper\_bound - lower\_bound + 1, 0\}$$

In case of an error, when the upper bound is less than the lower one, the number of iterations is zero.

In general, there is no direct possibility to deduce the maximum number of iterations directly from the loop condition, *e.g.*, when the bounds are expressed by run-time variables. In this case, the maximum number of repetitions must be specified by the programmer in the source code. Ideally, this can be achieved by extending the syntax of the loop statement:

```
for id=lb_expr to ub_expr with maxloop constant_expression do ...
```

Existing compilers, however, do not know such an extension of the programming language syntax as an option. Therefore, a specially constructed comment can be inserted into program code, so that execution time analysers can extract the information required from the comment and use it for evaluation:

```
for id= lb_expr to ub_expr do /*WCET:maxloop constant_expression*/...
```

In programming languages that allow one to define macros (*e.g.*, C/C++), a special dummy macro-declarations can be used:

```
#define WCET_MAXLOOP(X)
```

Then, much more readable code can be written, *e.g.*:

```
for id=lb_expr to ub_expr do WCET_MAXLOOP(constant_expression)...
```

These macros are evaluated by execution time analysers, but eliminated from the final code by a pre-processor.

Again, for precise analysis, the hidden administration time for loop preparation and index maintenance needs to be added. In addition, sometimes compilers considerably re-arrange the code loops. For example, the invariant part of the code (*i.e.*, the part not directly nor indirectly dependent on the loop counter), is removed from the loop body. Compilers may also unfold a couple of iterations into a sequence of statements, *etc.*

Other kinds of loops (*e.g.*, while or do-while statements) must also be bounded with maximum numbers of iterations.

The premature ending of a loop has no influence on the loop's maximum execution time. Further, there is no guarantee that the loop will not exceed the declared number of iterations. Therefore, care must be taken by the programmer or designer to assure compliance with the declarations. One possible solution to this problem is to use an additional counter. The counter is set to the maximum number of iterations prior to a loop's execution, and is decremented upon each repetition of the loop. If the counter turns zero, an exception is raised, as demonstrated by the code snippet in Figure 4.17.

```
guard_count := 100;
for i=a to b /* WCET:maxloop 100 */ do
     decrement(guard_count);
     if guard_count≤0 then
          raise ExceptionLoopExceedMaxIterations;
     end if;
     ...
end for;
```

Fig. 4.17. Raising exception on overflow of maximum iteration count

Parallel Execution

Some programming languages include syntactic elements for parallel execution of code parts on more basic levels; see Figure 4.18.

```
in parallel do
     s₁;
     s₂;
     ...
     sₙ;
end in parallel;
```

Fig. 4.18. Programming of parallel parts of code

From the semantics point of view, all statements s_i are executed in parallel. Each statement can be a compound one. It is presumed that there is no time-dependency between them. In reality, however, if only a single processor is available, only one alternative is executed at any time. If time slicing is used to mimic parallel execution, this means that the total execution time of the block is equal to the sum of the statements' execution times plus the overhead produced by the time slicing policy. Even in distributed execution environments with several processing elements, it is not practical to execute tightly related code on several processors in parallel.

4.2.3 Execution Time Analysis of Executable Code

The main difficulty in estimating worst-case execution time from source code is that compilers may re-arrange code for optimisation or other reasons. Because of this, the information collected by an automatic analyser could not be complete and correct. The complete picture of the actual execution of an application is contained in its executable code, only. Here, the application is represented as a stream of actual microprocessor instructions. If the execution times for the instructions were known, it would be easy to calculate the code's execution times.

By converting source code into microprocessor instructions, however, the information on program structure is lost. It is very difficult, if not impossible, to deduce the loops and alternatives in an original application from its machine code alone. Fortunately, during compilation, most compilers may also generate additional information about the composition of the code and, thus, retain structural information. They may associate specific addresses in the machine code with corresponding line numbers of the source code, they can gather information about the physical memory location of global and local variables, *etc.* This information is usually employed for debugging, but it may also be used for execution time analysis. Execution time analysers evaluate source code and extract the information about its program structure. After source code is compiled into machine code, analysers locate the specific application constructs in the compiled code and perform execution time analysis. The techniques for execution time analysis are similar to those used on the source code level, but more straightforward due to the simplicity of machine code instructions.

Nevertheless, there are some obstacles to implement this approach. First, a deep knowledge about machine instructions and their behaviour on the microprocessor is required. As stated above, modern microprocessor architectures can optimise instruction execution by executing them in parallel using different pipelining techniques. For this reason, the microprocessor manufacturers usually state only typical execution times of machine instructions. The next obstacle is that compilers may still re-arrange code in such a way that source code cannot be related one-to-one to machine code anymore. This can usually be avoided by turning off optimisation. After WCET analysis optimisation may be re-enabled. It is expected that optimised code will eventually be executed faster than non-optimised code, which means that the estimated execution times cannot be shorter than the actual ones. Executable code is usually stored in intermediate (re-locatable) form. Only during loading into main memory does the code assume its final form. Since this process is straightforward, it can easily be considered during execution time analysis.

4.2.4 Execution Time Analysis of Hardware Components

In embedded control systems, frequent interaction with hardware is common. It causes delays to program execution, and must be considered in the execution time of tasks. On distributed platforms, execution times are significantly influenced by the delays introduced by the communication infrastructure. Here, execution times must be evaluated on a case-by-case basis; there is no common method to evaluate them. If a target hardware is not yet built, the components' technical specifications can be studied. Manufacturers of hardware components usually give timing specifications for their products. Instead of maximum execution times, however, often typical or minimum execution times are only stated.

4.2.5 Direct Measurement of Execution Times

As discussed above, there are many obstacles in obtaining accurate and re-alistic execution times from static analysis of code. In most cases, execution time estimations turn out to be either too pessimistic when source code is executed, or too optimistic when machine code was analysed. In the first case, the cause is lack of knowledge of what code is actually produced for a specific statement. In the second case, the execution times supplied by the processor manufacturer for individual instructions usually apply for ideal sit-uations, when instructions are already in cache and neither stall in pipelines nor collisions on address and data buses are considered. It is, therefore, ob-vious that realistic and accurate execution times can only be determined by direct measurements on target (or equivalent) hardware platforms.

For this, a variety of measurement tools like software profilers, logic anal-ysers, or dedicated measurement devices can be employed. A profiler is a software tool that is mainly used to identify bottlenecks in code. It modifies code analysed by inserting special assertion instructions at all entry and exit points of procedures and tasks. Any time these points are reached, the addi-tional code is executed. It records the current time and the execution points' identifications. After a run, the execution history is analysed and the actual execution times are gathered.

The additional code introduces additional delays into program execution. To avoid significant interferences, the measurement points should be placed carefully. In case of WCET analysis, the problem of finding bottlenecks in code is not so relevant, because only task execution times are to be observed. To minimise interferences for tasks the insertions, only producing physical signals observable from the outside, should be placed at their entry and exit points. Logic analysers can observe these signals and record the durations between them. The process is very tedious and requires much human intervention.

The most straightforward way to determine execution times is by auto-mated measurement of hardware behaviour under control of a WCET anal-yser: the piece of code analysed is automatically furnished with signal gen-erators triggering the measurement. Then it is run and its execution time measured with a dedicated device as depicted in Figure 4.19. It consists of a high-resolution timer, which is started and stopped by writing into a reg-ister. The device needs to be connected to the processor bus and placed in the system memory address space. The instructions called probes, that access the register at the assigned address, start and stop the timer. The impact of the additional instructions to the total execution time of an analysed piece of code is negligible.

Best results in WCET estimation are achieved by integrating an analyser into a compiler. This approach allows both static analysis of application code by combining both source and machine code analysis, and proper direct mea-surement of execution times. As part of the compilation process, the latter

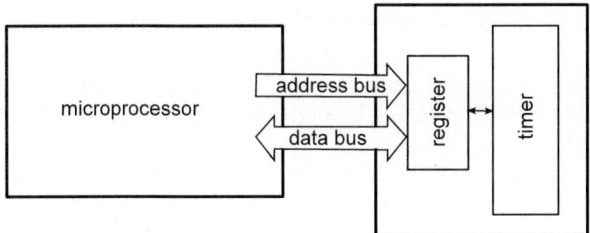

Fig. 4.19. Dedicated hardware device to measure execution time

gathers the internal structure of an application, which can be directly utilised by a timing analyser. This way, basic blocks are isolated.

A WCET analyser automatically loads the application code corresponding to the basic blocks onto the target hardware, instruments it with probes and runs it. The timer automatically records the execution time elapsed. These results are acquired by the analyser. Technically, this can be carried through by a real-time debugging support as already available in a number of microcontrollers. An example is the standardised JTAG Boundary Scan Mechanism [61]. A correspondingly equipped processor has dedicated hardware connectors *via* which it is possible to access its registers, control its behaviour, start or stop execution, set breakpoints, and even load or view memory contents. Debugging with the help of the JTAG mechanism is non-intrusive and does not influence the timing behaviour of code execution. Another similar example is the Background Debugging Mode (BDM) implemented in some Motorola microcontrollers.

This way, the analyser measures realistic execution times of basic block, which are then used in calculating the execution times of compound blocks as shown in the formulae of the previous section. In a recursive mode, finally the execution time of a complete program can be calculated.

Sometimes, better results are expected from directly subjecting compound blocks to measurement. Then some arrangements must be made; for example, all loops must artificially be re-arranged to execute the maximum numbers of times; all conditional jumps must be set to run always the longest alternatives, and so on. Furthermore, because tasks are analysed in isolation from each other, some system calls must be replaced by appropriate delays.

As a final note, even if direct measurement of execution times yields the best results, this must not be the only approach to temporal analysis. It is usually too late to begin with this at the final development stages. Moreover, measuring control applications requires that the applications already exist and that they can be run on hardware platforms at least similar to the final ones.

4.2.6 Programming Language Support for Temporal Predictability

Bounding Program Execution Times

The primary pre-requisite for execution time analysis is that the execution time of any statement is temporally bounded. Commands with non-deterministic execution times ought to be forbidden. If this should be impossible in specific cases, the corresponding program blocks must be temporally guarded, and time-out alternatives must be defined explicitly. An example is human intervention. An application may need the operator to enter a parameter for some control loop. The reaction time of the human operator cannot properly be predicted. The operator may even not reply at all and, thus, prolong the waiting indefinitely. It is a common sense that such kind of task cannot really be regarded as a hard real-time one.

There is a set of instructions common in multitasking embedded systems whose execution may take indefinite amounts of time, such as operating systems commands that are used for mutually exclusive access to resources or for task synchronisation. For example, critical sections are usually dealt with by some sort of locking mechanisms. Traditionally, waiting for synchronisation may take indefinitely long because of current conditions in a system: due to an error the locking synchroniser object may never be released, another task may always have precedence, or there may be a deadlock situation. In the end, of course, the deadline of the current task will be missed and the task will fail. The best way to prevent such situations is to employ appropriate structured synchronisation mechanisms. An example of them was shown in Figure 2.21 on Page 59.

Alternatively, locking mechanisms may be extended. For instance, instead of the `lock` command (*e.g.*, performing "wait for a semaphore") a `trylock` function may be used that tries to lock the synchroniser. If locking is possible, the function locks the synchronisation object; if not, an alternative statement sequence is executed. Another solution is to endow the lock with a timeout conditions. If locking is not successful during a predefined time interval, the `lock` statement has failed and the alternative set of statement is executed. All three cases are presented in Figure 4.20. In (a), the traditional lock/unlock mechanism is shown. By analyser instructions included in comments as described earlier, it is possible to assert an expected waiting time for WCET determination, but there is no guarantee that the given constraint will actually be meet. In (b), the `trylock` function is used to execute two alternative blocks of statements depending on whether the lock was successful or not. Similarly, it can be used for condition checking, *e.g.*, to re-try locking several times within a while-loop statement with the `trylock` term as exit condition. In (c), the `lock` statement is incorporated in a try-catch block. If locking is not successful within a predefined time interval, the `timeout` exception is raised to be handled by the exception handling mechanism.

```
(a)                    (b)                                (c)
lock sync-object       if trylock sync-object then        try
...                        /*the lock was successful*/        lock sync-object with time-exp
unlock sync-object         ...                                /* the lock was successful*/
                           unlock                             ...
                       else                                   unlock sync-object
                           /*the lock was unsuccessful*/  catch
                           ...                                on timeout
                       end if;                                /*the lock was unsuccessful*/
                                                              ...
                                                          end try;
```

Fig. 4.20. Three examples of lock/unlock mechanisms: (a) with indefinite waiting when lock is not successful, (b) with trylock mechanism, and (c) and with timeout condition

In all cases described above, the lock statement was followed by an unlock statement. This is not strictly necessary. For example, one task can lock a synchronisation object and another may unlock it. However, it is always better to use a structured form of locking. This way, the time during which a task is within a critical section can be determined, and it can be taken into account in course of schedulability analysis.

Bounding Loop Iterations

To perform execution time analysis, the maximum number of repetitions of iteration statements must be known. For this, one of the techniques described in Section 4.2.2 (*i.e.*, extending the syntax of the loop statement or including compiler and analyser instructions in strictly defined comments) should be used declaring constant compile-time expressions to restrict the number of loop iterations. Programmers must also assure that the limits set will actually never be violated. As the most secure method, only for-loops with fixed numbers of iterations (with potential earlier exits) should be allowed.

Renouncing or Bounding Recursion

For similar reasons as in the case of loops, the use of recursion can cause temporally unpredictable behaviour. Similar syntactic constructs as for loops can be used to limit recursions. Some forms of recursion, however, are difficult to analyse automatically, *e.g.*, a procedure may call another one and this recursively calls the former one. Furthermore, unbounded recursion can result in severe memory management problems at run-time. Any procedure call (recursive or not) makes use of stack memory. Even in systems with an abundance of memory available, recursion may cause stack overflow. As a solution, recursion should be converted into iteration which is, theoretically, always possible.

Preventing Dynamic Data Structures and Memory Management

The use of dynamic data structures is highly data dependent at run-time. For example, the execution times of search algorithms on graphs represented by dynamic data structures depend on the number of graph nodes. Dynamic memory allocation algorithms are already by themselves data dependent. Further, successive allocation and de-allocation of memory can generate memory fragmentation, *i.e.*, a large number of small free memory blocks emerges, but none of the chunks is large enough to suit larger memory requests.

Several modern programming languages (*e.g.*, Java) employ memory management strategies where allocated memory is not (or cannot be) freed even if the associated structure is out of scope and destroyed. Only when there is not enough memory available (or periodically), a so-called garbage collector is invoked trying to reclaim and pack together the free memory available. This strategy causes occasional non-deterministic actions that cannot be predicted during timing analysis. Consequently, one should rely on static data structures, only, with *a priori* observed memory requirements. For example, data arrays with fixed dimensions can be used to represent graphs. The search algorithms on graphs may still be used, however, as the number of array elements is limited, so is the maximum number of steps performed by an algorithm.

Prohibiting GOTO Statements

The execution times of GOTO statements *per se* can easily be determined. The use of them, however, can result in unstructured and hardly manageable code, which cannot be analysed automatically. Jumps that commit early exit from inner blocks (*e.g.*, EXIT or RETURN), and branches that immediately initiate the next iterations of loops are allowed. Such statements eventually reduce total execute time while leaving worst-case situations unchanged.

Explicit Assertion of Execution Time

Frequently, because of the nature of a program, blind automatic estimation may yield very pessimistic execution times. An example for this is shown in Figure 4.21, where the execution times of two subsequent loops should not be added. If the first loop is executed more often, the second loop is executed less times, and *vice versa*. To resolve this problem, additional information about program execution should be specified by the programmer. This can be achieved by adding new constructs (pragmas) to the language syntax as proposed earlier. Thus, in the example it can be noted that the sum of the repetitions of both loops is constant (here always 100).

Since constructs as mentioned above require complex analysis and are not feasible for all situations, it is often more appropriate to set execution times explicitly. Based on trustworthy measurement, detailed analysis of program behaviour, experience, re-use, *etc.*, an execution time may explicitly be asserted by the system developer, and execution time analysis is overridden. To guarantee that the actual execution time will not be longer than declared, however, the pertaining block must be guarded by time-out control, and a time-out action must be specified, as shown in Figure 4.22.

```
for i=1 to 100-n with maxloop 100 do
  ...
end for;

for i=100-n+1 to 100 with maxloop 100 do
  ...
end for;
```

Fig. 4.21. Sequence of two loops where automatic execution time analysis yields too pessimistic results

```
try
    with timeout time-exp do
    ...
    end with;
catch
  on timeout
  ...
end try;
```

Fig. 4.22. Guarding of a block with time-out control

Use of Mass Storage Devices, Expert System Libraries and Data Bases

Mass storage and asynchronous input/output devices as used in conventional computer systems are not suitable for embedded hard real-time systems. There is, for instance, a huge difference between minimum and maximum access times for magnetic disk storage leading to unrealistic estimations of execution times. Similarly, all software algorithms and methods with a wide variety of execution times should also be prevented by incorporating the utilisation of expert systems, machine-learning strategies, heuristic methods, sophisticated data bases, *etc.* into soft real-time tasks.

4.2.7 Schedulability Analysis

Schedulability analysis is a process of formally verifying whether a control system will meet its specified timing requirements under any circumstances. It is based on given deadlines, execution times of tasks, and other timing specifications. In general, schedulability analysis is a problem whose complexity grows exponentially with the conditions of task execution. To be able to prove feasibility, usually some additional constraints must be imposed on system behaviour. For example, assume that a system is to perform a task scheduled for an external event (*i.e.*, it is activated any time the external even occurs). If the time period between two occurrences of the external event is long enough, the system can perform the corresponding task easily. However, if the period between the event occurrences is decreased, a new event may arrive before the previous one is processed. If there are other tasks with higher priority in the system, the danger of overrun is even higher.

It is, however, often unreasonable to assume that the period of an event's repetition can be arbitrarily short. For example, if an event is based on data from a temperature sensor, the process of reading sensor data includes analogue-to-digital conversion that takes some time. Furthermore, it is unreasonable to check inputs too frequently which exhibit high persistence for physical reasons (such as temperature of a large mass). Schedulability analysis can render good results only on the basis of reasonable estimations.

Another difficulty in schedulability analysis originates from the assumption that each event (synchronous or asynchronous) may arrive at any time. If time is considered as a continuous quantity, the number of possible scenarios of event arrivals is infinite, and analysis becomes very difficult. A common solution to this problem is to discretise time. Now, changes (or reactions) may happen only between two time units. By this, the number of possible combinations to be observed is still large, but (hopefully) manageable.

All assumptions made for schedulability analysis must also be assured in real system behaviour. For instance, the assumption that an event will occur with a specified maximum frequency, only, must be supported by hardware means. In particular, the observation of discrete time must be established. This execution model is used by the architecture proposed in Section 3.2. The same model is employed in programmable logic controllers, where reading of inputs, execution of control algorithms, and setting of outputs occur in statically defined constant cycles.

An example of a programming language that models the execution of applications with discrete time units is Real-Time Euclid [50, 69], defining the basic time unit by

```
RealTimeUnit(time_in_seconds)
```

In any time unit, only one activity may be executed. If the task ends earlier, the rest of the time slot is wasted. Real-Time Euclid was designed specially to support schedulability analysis, as a prototype laboratory language. It includes other properties to enable temporally predictable execution (*e.g.*, it has no dynamic data structures nor recursion, loops are bounded, blocking time of synchronisation is maximised). For *a priori* schedulability analysis, a method called frame superimposition is utilised: after every non-interruptible task segment all combinations of successive continuations of possible non-interruptible task segment are taken in consideration. It is demanded that all deadlines are met — even in the most unlikely case. Evidently, the method's complexity is NP complete and the results are extremely pessimistic. Thus, this analysis is suitable for simple cases, only. Nevertheless, the schedulability of asynchronous task sets can be proven.

4.3 Object-orientation and Embedded Systems

In the 1990s, the paradigm of object-orientation became more and more popular, although its concepts were first introduced in the 1960s into programming languages like Simula and Smalltalk. From the object-oriented (OO) approach it is expected to improve quality and efficiency of building software applications by incorporating several programming concepts. The first concept, modularity, groups together related data structures and operations on them into a single entity, called class. This is similar to the technique structuring programs into separate modules. A module can maintain a set of data elements and a set of operations which are (usually) related to this data. While there can exist only one module of a kind within an application, a single class may be used to instantiate several data elements (or objects) that share the same characteristics.

Another aspect of classes is that they can be considered as natural extensions of classical compound data structures. A compound data structure comprises a set of data elements representing related pieces of information. Based on a structure definition, several data entities (*e.g.*, variables) can be instantiated, which share the same internal structure but own individual copies of data values. Access to the individual data elements within an entity is only possible through references to them. In addition to data elements, classes also define sets of operations (called methods) that may be used upon them. Analogue to accessing data elements, these operations can be executed on the objects instantiated by a class, only

Modularity also provides for another important programming technique, *viz.*, information hiding or encapsulation. For the user of a program module or class it is only important to know which data elements and which methods are available. Knowledge of the methods' actual implementation is usually not necessary. In programming modules, information hiding is achieved by dividing modules into interface and implementation parts. In class definitions this is achieved by marking some class elements as private and the others as public. The former can be accessed only inside the implementation of class methods.

Another important concept characterising the OO approach is inheritance. It allows to extend the functionality of a class by adding additional data elements and/or additional methods in the course of defining a new class. At the same time, the derived class keeps (inherits) all functionality of the base class. This approach significantly enhances code re-usability. Programmers do not need to have the source code of the original classes' definition and implementation; instead, a compiled version of the class definition from some program library can be used. If necessary, the derived class may replace (override) some functionality of the base class (*i.e.*, the implementation of a specific operation is replaced by a new one). Now, when such a method is called in an object, the code actually executed depends on the class from which it was instantiated. This property is called polymorphism and represents another important concept of OO programming.

Some programming languages support multiple inheritances, *i.e.*, a derived class inherits properties from more than one base class. Its class definition may also define special kinds of methods, constructors and destructors. The former are executed automatically when the object is instantiated, and are used to initialise its data structures. The latter are called immediately before the object is destroyed or when it leaves scope.

4.3.1 Difficulties of Introducing Object-orientation to Embedded Real-time Systems

There are several obstacles to the utilisation of object-orientation in embedded real-time systems. For example, the use of polymorphic methods may render proper execution time analysis difficult. Since the decision which code will be used when calling a method is only resolved at run-time, *a priori* WCET analysis is lacking important information. One solution to this problem is to consider always the implementation with the longest execution time which, however, may yield too pessimistic estimations.

Another difficulty is merging the OO approach with parallel activities in real-time systems. For example, two method calls of the same object could be executed at the same time by two different tasks. Both of them may arbitrarily change the data values in the objects in an uncontrolled way. The basic OO approach does not provide a proper solution. Therefore, inside the methods a synchronisation mechanism must usually be provided by the programmer. One approach would be to implement the objects in a way similar to monitors (*i.e.*, only one method of an object can be executed at any time). However, the standard programming languages do not provide for this.

A further important issue for successfully applying the OO paradigm in the real-time systems domain are human factors. Application developers of embedded systems are usually engineers with broad knowledge in a problem domain, but with limited knowledge on object-oriented software development. This methodology is mainly used by computer specialists, and requires different approaches and a different way of thinking. Even when an object-oriented programming language is employed by the engineers, often the features of object-orientation are omitted.

4.3.2 Integration of Objects into Distributed Embedded Systems

In object-oriented design, system elements are considered as objects, collected in classes. In embedded systems development, traditionally the structured approach is used. It may be beneficial, however, to represent traditional elements of real-time systems as objects, thus introducing the object-oriented approach.

Task Considered as Object

When implemented in programming languages, tasks are often considered as a special kind of procedures. As was noted previously, in many programming languages tasks are not supported explicitly; instead, operating system calls are used. On the other hand, objects are traditionally considered as extension to static data types (like structures in C++). Data elements of these structures are upgraded with local procedures (methods). Objects do not have direct control over their execution. If parallel execution, synchronisation *etc.* is required, operating system calls must be used.

One possibility to implement a task as an object is to construct a so-called wrapper class around system routines. For example, based on the POSIX standard introduced previously, a class Thread can be specified as shown in Figure 4.23.

```
class Thread
{
private:
  int thread_handle;

  Run()
  {
    exit_status = Execute();
    pthread_exit(exit_status);
  }
public:
  Thread(thread_attributes) /* constructor */
  {
      ...
      pthread_create(thread_handle,thread_attributes,
                     AddressOf(Run),...);
      ...
  }
  virtual int Execute() { return 0 };
}
```

Fig. 4.23. Specification of the thread in POSIX

The class Thread has a constructor that creates, during object instantiation, a POSIX thread using given attributes. The result of the call, a thread handle, is assigned to a private variable and can be used in subsequent system calls. The body of the thread is defined with virtual function Execute. It returns the thread's exit status. This function is called indirectly by supporting method Run which is passed as an actual parameter with the POSIX thread creation. By this, it is possible to implicitly perform **pthread_exit** (and any other clean-up activities) when the thread code is finished. To use this class, a new class must de derived and the Execute routine overridden with the actual code; see Figure 4.24.

```
MyThread class(Thread)
{
public:
  override int Execute()
  {
    ...
  }
}
```

Fig. 4.24. Overriding of a virtual routine with actual code

A better solution would be if operating systems treated tasks as objects. Then, all particularities of task handling would be integrated as class properties. A hypothetical example for this reads as shown in Figure 4.25

```
Task class {
private:
  t_context TaskContext;
  ...
protected:
public:
  Thread(t_duration Deadline); /* constructor */
  virtual void Activate(t_schedule Schedule);
  virtual void Terminate();
  virtual void Suspend();
  ...
  virtual void Execute() {  };
};
```

Fig. 4.25. Hypothetical example of a task as object

Task declarations have several private variables (*e.g.*, internal context, current state) and several private methods. They all serve the needs of the operating system and are used as primitives for other method implementations. Traditional tasking operations (*e.g.*, Activate, Terminate, *etc.*) are defined as public methods. In addition, some task properties may be interesting to programmers. They are available through several public functions. Actual execution of these methods depends on the specific implementation of a control system. In a single processor system, methods are executed directly by the operating system. On a distributed system with a separate kernel processor, each command is translated into a message and sent to the kernel processor through the communication network.

For application developers, only the public parts of class interfaces are important. Most public methods are declared as virtual. By their re-implementation, additional logic may be provided. Similar to the previous example, the main body of a task is not known in advance and must be overridden as illustrated in Figure 4.26.

```
MyTaskType class(Task)
{
public:
override void Execute()
  {
    . . .
  }
}

/* declaration of the task */
MyTaskType MyTask(10 MS);

/* task activation */
MyTask.Activate(NOW);
```

Fig. 4.26. Overriding of a routine with actual code, again

The real advantage of the object-oriented approach becomes effective when general tasks are classified into different groups. For example, many tasks in real-time systems are periodic, with their operation triggered in regular intervals. For this group of tasks a new class can be derived from the general one; see Figure 4.27

```
PeriodicTask class(Task) {
private:
  t_duration period;
  . . .
public:
  . . .
  virtual PeriodicTask(t_duration Period); /* constructor */
  . . .
};
```

Fig. 4.27. Deriving a new class for a group of tasks

Tasks of this class have an additional property: the activation period. Several methods of the general task class are simplified or hidden. With this classification of tasks, programming code can be more expressive.

Other Elements of Real-time Systems as Objects

Similar to tasks, other real-time constructs can be represented as objects, as shown in Figure 4.28 for semaphores.

```
Semaphore class {
private:
  ...
public:
  void Semaphore(int InitialCount); /* constructor */
  void Signal();
  void Wait();
}

/* declaration of the semaphore */
Semaphore S(1);

/* usage of semaphore */
S.Wait();
...
S.Signal();
```

Fig. 4.28. Semaphore as an object

Again, the method implementations are hidden. Semaphores and similar programming constructs are well defined, and there is usually no need to derive new classes from them.

As in the case of tasks, classes can be better utilised if similar components can be arranged into consistent groups. The most likely candidates for grouping are hardware components. There is a wide range of input/output devices and similar components with corresponding methods to access and control them. Here programming particularities are combined with construction specifications. Some properties of these objects are only used in hardware implementation (*e.g.*, an interrupt number or addresses of registers in peripheral interfaces). These properties can be marked as specification attributes and are eliminated by the compiler. For example, a class for a general memory-mapped I/O device can be declared as shown in Figure 4.29.

Prevention and Handling of Deadline Violations

As discussed above, even if formal schedulability analysis is performed properly, sometimes task deadlines may not be meet due to transient system overloads, hardware failures, *etc.* In some cases, the operating system can detect

```
MMIODevice class {
public:
    virtual void MMIODevice(int PhysicalAddress); /*Constructor*/
    virtual void SetValue(int Value);
    virtual int  GetValue();
}

/* declaration of the devices */
MMIODevice AD(0xCCCC001);
MMIODevice DA(0xCCCC003);

/* usage of the devices */
ad_val = AD.GetValue();
...
DA.SetValue(da_val);
```

Fig. 4.29. I/O Device as an object

such a situation early enough to be able to deal with it. If such an error is detected, some tasks can be replaced with alternatives having shorter execution times.

Since deadline prevention is under the control of the operating system, it must know all possible execution alternatives of the application tasks in advance. Here, object-oriented representation of tasks can be utilised. In addition to a task's main execution procedure, as shown in Figure 4.30 several alternative procedures can be declared.

```
SafeTask class(Task) {
public:
    ...
  virtual void MainExecutive();
  virtual void AlternativeExecutive1();
  virtual void AlternativeExecutive2();
};
```

Fig. 4.30. Alternative implementations of task's main execution procedure

In this example, two additional procedures are provided to prevent deadline violations. During WCET analysis, the execution times of all three alternatives are determined. The operating system must be aware of the alternatives (*e.g.*, by using different tasking primitives inside the SafeTask class declaration). During normal operation, MainExecutive is executed. If an imminent deadline miss is detected, the operating system uses the alternatives' execution times to determine if a different task body can feasibly be executed.

Software Redundancy

To increase the reliability of and to cope with catastrophic failures in distributed systems, redundancy is employed. With this, a failure in one system part does not jeopardise correct system operation as a whole. It is essential for safety critical-systems to have redundant hardware components (*i.e.*, sensors, actuators, communication channels, processor boards *etc.*).

Redundancy in slightly modified form can also be applied to software. In distributed systems, routines can be run on several processors in parallel. Even if there is an error on one of the processors, the results from the others can be used. Furthermore, to counteract logical errors introduced by flawed specifications, diverse implementations of one routine may be used and run at the same time. At the end, results from different replicas are compared and the best one is used in further computation. The alternative routines can be implemented with the object-oriented approach in a similar way as were the diverse tasks earlier in this chapter.

Trying to implement this feature, however, several difficulties are encountered. For obvious reasons, the alternative routines cannot directly change internal system states (*i.e.*, output devices, shared variables *etc.*). Only pure program functions (*i.e.*, procedures without physical effects) can be used effectively. These problems are even more pronounced when objects are considered. Any object owns a set of internal variables. When several copies of the methods of a specific object are executed, these variables cannot be changed simultaneously unless each method has its own copy. If communication channels are slow, this is too time-consuming. Using a copy of the object as a whole seems to be the only feasible approach. The same or similar objects should run on separate processors. Direct modification of shared variables and states of input/output devices is forbidden. All results must be forwarded to a separate task that compares them and selects from them the final result to change the system state or to send appropriate commands to peripherals. This kind of redundancy is hard to automate. Usually it is under direct control of the developer. Also temporal analysis of such systems is quite complex.

4.4 Survey of Programming Languages for Embedded Systems

At the dawn of computerised process control, only assembly languages were used, supporting rather simple processors, and optimising the utilisation of limited memory and the access to low-level peripherals. While assembly languages are still useful in building real-time applications with tight timing or memory constraints on basic microcontrollers, high-level general-purpose programming languages prevail as means for programming. They boost productivity, introduce programming discipline, and foster the use of rigorous methods. Some high-level programming languages have been adapted to program embedded real-time systems. In most cases, however, real-time functionality is available to programmers only through operating system calls. This category of programming languages is referred to as real-time implementation languages. Relatively early in the 1960s, special-purpose real-time programming languages emerged. It is interesting to note that they are still not widely used in spite of their advantages and of the problems caused by using implementation languages to program embedded systems. In this survey, selected languages will briefly be discussed. A more comprehensive survey on languages for real-time programming can be found in [49, 50, 11, 21].

One of the languages including most of the properties desired for developing real-time programs is Real-Time Euclid [69, 50], which was briefly introduced in Section 4.2.7. Its objective is to produce temporally deterministic and predictable code in order to support schedulability analysis. This language is, unfortunately, only of academic importance as a prototype, proposing solutions to open issues of real-time programming. No compiler for this language is commercially available.

In large-scale embedded systems for automation of industrial processes, implementations on the basis of programmable logic controllers (PLC) prevail, which use a different programming paradigm than traditional programming languages. Their programming will be briefly dealt with at the end of this survey.

4.4.1 Assembly Language

Assembly languages provide for the most direct interaction with hardware and the most transparent temporal behaviour. On the other hand, they lack any other property needed for programming real-time systems or for any kind of large-scale programming. They are unstructured, hard to learn, ineffective for human use, and error-prone. The programs are not portable. For tasking and synchronisation they are fully dependent on operating system calls.

However, programming in assembly languages cannot always be avoided. For certain microcontrollers and for most custom-built hardware systems no high-level language compilers are available. Furthermore, for devices with limited processing capabilities and limited resources, the use of high-level programming languages is not feasible, because it often yields inefficient code. With the availability of memory and processing speed becoming much less restrictive on most platforms for embedded systems, however, the use of assembly languages should be discouraged.

Nevertheless, assembly languages can still provide for low-level hardware access to input/output interfaces, implementation of interrupt routines, or management of devices with direct memory access. Assembly code is still used in those parts of control applications where tight timing constraints need to be met, and to remove bottlenecks in programs. Some compilers of high-level languages allow insertion of fragments of assembly code directly into application source text. If this is not supported, routines written in assembly code can be compiled separately and linked to other object code later.

4.4.2 General-purpose Programming Languages

FORTRAN is the earliest commercially successful high-level programming language developed in the 1950s [4], and is especially useful and still popular for programming in technical fields. As such, it was the obvious choice for use in process control. Several versions of FORTRAN exist; each new version introduced new syntactic and semantic elements into the language, maintaining backward compatibility. There is a large code basis for embedded systems written in FORTRAN still running today, especially for military purposes. Nowadays, FORTRAN is used as a development language for some applications because of its advantages with respect to human productivity. In embedded systems applications, it was in most cases replaced by C.

Compilers for FORTRAN are known to produce very efficient code due to the language's straightforward syntax and semantics. However, FORTRAN requires interfacing with assembly code to support real-time functionality like interrupt handling, low-level input/output device interactions, or scheduling. The first versions of FORTRAN did not even include bit-manipulation routines.

In the 1970s, Industrial Real-Time FORTRAN (IRTF) [70] was developed. Apart from providing for process, device, and I/O control, it also included a set of bit-manipulation functions, which were eventually incorporated into Military Standard ANSI X3.9-1978 for FORTRAN MIL-STD-1753 for FORTRAN-77, and later into FORTRAN-90.

The programming language **C** was introduced in the early 1970s to implement system software instead of assembly language. Today it is one of the most popular programming languages for embedded systems. It requires a simple compiler and only small run-time support. In 1990, the programming language C was standardised as ANSI C, and most C code written these days is based on this standard.

On the other hand, C is not very suitable to develop dependable and real-time applications. It has weak type checking, no checking of array bounds, lacks structured exception handling, its memory management is under programmer control, its support for dealing with time is limited, for tasking and synchronisation operations it relies on operating systems, *etc.* Because of these limitations of C, several new dialects and extensions emerged to make it more suitable for dependable real-time applications. For example, the Cyclone programming language was devised to avoid buffer overflows and other vulnerabilities of standard C.

Another approach to cope with the limitations of C was to define a set of guidelines and rules for safe programming that do not extend the syntax of the language. One of the most well-known examples for this is MISRA C [84], a software development standard conceived by the Motor Industry Software

Reliability Association for the implementation of reliable embedded systems programmed in ANSI C. Some rules of MISRA C can be checked by a compiler; other more general ones influence the development process as a whole.

Several dialects of C attempt to introduce object-orientation into the language. One of them, "C with Classes", became C++, a widely used general-purpose programming language today. Besides stricter type checking and structured exception handling, however, C++ does not introduce enhancements valuable for embedded real-time applications. Therefore, similar to C, many dialects of C++ have been introduced to make it more suitable in the field. One example is Embedded C++ [33] that preserves the basic features of object-orientation in C++ while renouncing some others not desirable for embedded systems like multiple inheritance, virtual base classes, or templates. The language is, however, oriented at 32-bit RISC processors, and falls behind by not offering exception handling.

Java was introduced in the early 1990s with the original goal of making object-oriented programming suitable for embedded systems. Java is similar to C++, but introduces some restrictions that are to make applications more reliable. For instance, pointers are replaced by implicit references to objects, all references are type-checked, and array references are subject to strict boundary checking. The main feature of Java is platform independence. Java code is first translated into an intermediate code for a hypothetical "Java Virtual Machine" (JVM). This code is then interpreted on the target system. Alternatively, prior to its first execution, intermediate code can be translated into the native object code of the target processor.

The most distinctive property of the Java programming language is its automatic memory management. Any time an object is created, an adequate amount of memory is allocated. When the object turns obsolete, however, this part of memory is not freed automatically. It remains inaccessible or "garbage". Only when there is no more free memory, or periodically, a so-called "garbage-collector" is called which collects any unused memory references, and frees and defragments memory. This automatic memory management is the main reason why standard Java is only conditionally appropriate for hard real-time applications. It is undetermined at what time the garbage collector is executed, and how much time it needs to finish its work. Moreover, the following characteristics render Java unsuitable for real-time programming: it does not support direct access to low-level input/output peripherals, memory access is restricted in some implementations to a set of resources managed by the execution environment (so-called "sandbox"), and there are no direct provisions for real-time tasking and synchronisation.

Because of these limitations, different extensions to standard Java were introduced to support real-time programming. The most widely used extension is based on the so-called Real-Time Specification for Java (RTSJ) [96], which is no syntactic extension to Java, but a set of programming interfaces (classes) and behavioural specifications that allow standard Java programs to be run in a real-time fashion. An RTSJ application is compiled with a standard Java compiler and executes on a special RTSJ virtual machine. The main objective of RTSJ is to be backward compatible and independent of particular development environments. Its highest priority is predictable execution. To this end, it defines seven "areas of enhancement" where the specification of standard Java is extended. The first one considers automatic memory management. By RTSJ, a program is allowed to allocate memory outside the control of the standard garbage-collected heap. Further, RTSJ introduces a class for real-time tasks and supporting classes for enhanced scheduling, synchronisation, and resource sharing. The scheduling policy is not restricted to any particular implementation, and can be adjusted to the actual operating system used by the target system. Additional classes implement interrupt and signal handling, *etc.* Finally, the class "RawMemoryAccess" allows one to access memory directly.

4.4.3 Special-purpose Real-time Programming Languages

HAL/S is a high-level programming language introduced by NASA in the late 1960s to meet real-time programming needs and to replace programming in assembly languages. In contrast to other languages, the HAL/S notation uses subscripts and superscripts to make the language more readable for engineers. It includes TASK program blocks, and simple priority-based scheduling constructs like SCHEDULE, WAIT and TERMINATE. With the SCHEDULE statement a task activation can be scheduled either periodically or upon arrival of an interrupt. One of the objectives of HAL/S was high reliability. A program should produce correct results on all possible inputs. For this, certain high-order language constructs (like GOTOs) are eliminated from the language or formally restricted, memory requirements of applications can be determined *a priori*, *etc.* HAL/S was used to program the on-board computers of the Space Shuttle and some other spacecraft. In the mid-1980s, however, NASA gradually replaced HAL/S by Ada.

JOVIAL was developed for large-scale real-time programming in the late 1950s [99]. It was mainly used to write software for the electronics of military aircraft. In 1973, it was standardised as MIL-STD-1589, and later revised with MIL-STD-1589C. Today, it is mainly used to update and maintain older application programs embedded in on-board avionics, tactical and strategic missiles, ammunition, and space systems [68]. From JOVIAL programs code is generated for a 16-bit microprocessor standardised in MIL-STD-1750A. It can, however, be used for other target platforms as well. JOVIAL supports straightforward compilation. It allows inserts of assembly code and direct access to hardware. For distributed computing, JOVIAL uses a common runtime repository (so-called communication pool or COMMPOL) which can contain shared variables, records or tables. On the other hand, it has only rudimentary support for parallel execution and exception handling.

Modula-2 was developed in the early 1980s as the successor to the programming language Modula that was never released as a commercial product

[116, 85]. It is based on the programming language Pascal, is well structured, and imposes strict type checking. The name is derived from the fact that Modula supports partitioning of applications into modules with separate definition and implementation parts. This facilitates large-scale programming and better machine-independence (*i.e.*, the hardware specific part of a code is incorporated in the implementation part of a module; for a new target system only this part is replaced by a new one while its interface remains unchanged).

The original Modula includes the concept of processes and their synchronisation with signals. This is replaced in Modula-2 by the lower-level notion of so-called coroutines, which are autonomous parts of code similar to tasks. Parallel execution of them, however, is under control of the programmer, who must explicitly transfer the program flow from one coroutine to another. For other system facilities, operating system calls must be used.

To eliminate some drawbacks of Modula-2, Modula-3 was introduced in the late 1980s, but was never widely used. Among other features, Modula-3 offers properly structured exception handling, objects, and automatic garbage collection (a concept later taken over by Java). Instead of coroutines, lightweight threads are provided.

PORTAL was developed in the mid-1970s for the implementation of real-time process-control applications [98]. It has many features of programming languages well-suited for the development of dependable real-time systems. Nevertheless, its commercial usage today is very limited. Similar to Modula, its syntax is based on the programming language Pascal. For parallel activities, PORTAL offers processes, semaphores, and monitors. Waiting for synchronisation can be limited with timeout clauses. To avoid memory overflows, the compiler checks maximum stack length for each process, taking all routine calls into account. Memory is then allocated for all processes in advance. PORTAL allows direct access to the hardware. Interrupts are inherently transformed into signals that can serve as activation conditions for specific processes. As a drawback, PORTAL lacks structured exception handling.

Ada is a procedural programming language that was designed under a contract of the United States Department of Defense during the early 1980s [1]. The main intention was to replace diverse programming languages used in different military embedded real-time applications by a single one. For this, a series of requirements documents was prepared which eventually evolved into the Ada83 specification. This includes a rich set of features, which requires, as a drawback, a powerful compiler and extensive system software support. Consequently, the early Ada applications were not well suited for platforms with limited processing resources. A later version of Ada (Ada95, [65, 2]) changed this. Instead of a large number of features, only a set of core features must be implemented by Ada compiler. Development systems may implement additional sets of features defined by so-called annexes. For example, Annex D defines features for real-time systems, Annex E features for distributed systems, and Annex H security and safety features. With further annexes, Ada95 broadens its usage for other kinds of applications (system programming, information systems).

Ada is a structured and strongly typed language. It uses packages to partition applications into modules. Comparable to Modula, each package has a public interface part and a protected implementation part. Parallel activities are supported by the task concept. Similar to packages, the specifications and implementations of tasks are separated. For synchronisation, the rendezvous mechanism (see Section 2.3.5) is used. Ada allows structured low-level access to input/output interfaces, and has structured exception handling. On the other hand, the features provided for embedded real-time systems are somewhat limited: there are only simple tasking operations, just priority-based scheduling, and rather rudimentary time-related operations.

PEARL (Process and Experiment Automation Real-time Language) [29, 30, 92] is a high-level language which allows one to write multitasking and real-time programs in a comfortable and widely hardware-independent way. Since the language syntax explicitly states most particularities of real-time systems, it is better suited for real-time programming than Ada. Therefore, examples expressed in PEARL syntax are used throughout this book. There are four standardised versions of PEARL. The last one, PEARL90, was introduced in 1998.

A program in PEARL consists of a so-called "system part" and several "problem parts". The system part isolates particularities of the hardware platform from the algorithm, which is defined in the problem parts. This allows for better hardware independence. On a new platform, only the system part must be replaced. To support time, PEARL explicitly defines two time related-data types, `CLOCK` for time instants and `DURATION` for time intervals, with corresponding arithmetic operators. It allows both high-level and low-level interaction with input-output devices by using so-called data stations (or DATIONs). Data stations represent virtual input/output devices, and constitute a uniform approach to communicate with various hardware components.

To model parallel activities, PEARL employs the task concept. Its support for tasking operations is very rich. Tasks can be scheduled for activation upon a wide range of temporal conditions, or on occurrences of interrupts. For synchronisation, general semaphores and bolts are used. PEARL is problem-oriented and can easily be used by engineers. The written code resembles plain natural language.

A specialised version of the language, PEARL for Distributed Systems [30], is virtually the only standardised programming language suitable to program distributed systems. It allows one to describe physical platform topologies, and supports the declaration of different hardware and software configurations. Which configuration will be used depends on a system's current operation mode. This directly supports different re-configuration scenarios for different cases of faults.

As drawbacks, PEARL lacks properly structured exception handling, uses unstructured synchronisation mechanisms, implies priority-based scheduling, and has limited provision for temporal predictability. A number of suggestions to improve the language has been made. In [109, 51] several extensions of PEARL have been proposed to support different levels of dependability. Propositions to extend the PEARL syntax with object-oriented facilities were described in [42, 113].

4.4.4 Languages for Programmable Logic Controllers

Programmable logic controllers or PLCs are dedicated microcomputers mainly used in the automation of industrial processes. They are meant to operate in harsh conditions. A PLC application executes in a cyclic fashion: first, the inputs are acquired, then processing is performed and, finally, the outputs are produced — all within a single cycle. After that, execution pauses until the start of the next cycle. In 1993, the international standard IEC 61131-3 [56] took effect defining four different but equivalent programming languages for PLCs: ladder diagrams, function block diagrams, instruction list, and structured text. Programming of PLC applications in these languages is problem-oriented, and portrays the control engineering way of thinking.

Because of their simple execution model, PLCs operate in real time; the reaction time is bound to the period of one execution cycle. On the other hand, because of this execution model, more complex control applications may not be implemented with PLCs. It is possible to skip execution of any function block in a specific execution cycle. Therefore, it is possible to implement some rudimentary timing conditions, *e.g.*, a function block may be executed every second, third, *etc.* execution cycle.

The paradigm of PLCs and its execution model is widely used in control engineering. Applications are described as sets of interconnected function blocks, which are executed in cyclic fashion. Each function block is described either mathematically or by some simple programming code. Applications can be simulated and the results analysed on their host systems. One of the widely used development tools for this kind of applications is Simulink®.

Part II

Implementation

Case Studies on the Implementation of Design Guidelines for Embedded Control Systems

In Part 1 of this book, guidelines to design embedded systems have been given. Starting with the concepts of real-time operation and principles of fault tolerance, ranging over matters of tasking including synchronisation and scheduling, more specific topics were elaborated, namely, architectural and programming issues. The intention of Part 2 is to show how the above guidelines can be used in practical implementations.

During the last 15 years, several laboratory prototypes have been implemented, tested and evaluated. The prototypes were based on different hardware platforms, starting with the first one dating back to the early 1990s. It was based on diverse microprocessor boards built around, for that time relatively powerful, INMOS transputers T425 and Motorola microcontrollers MC68306 and MC68307. The implementation was presented in [18]. Its interesting characteristic are point-to-point connections between the kernel- and task processors based on the simple transputer's serial links. Peripheral interfaces resided at a serial local bus driven with the I^2C protocol. The architecture of this platform has briefly been presented in Section 3.3.1 and is sketched in Figure 3.5. The platform was renounced due to technological advances and insufficient flexibility. From this prototype, several others evolved, featuring more flexible distributed control and gradually improved fault tolerance.

The latest hardware platform is, in more detail, presented in Chapter 5 of this part. It has been designed in the course of a research project [62, 82], carried out within the European Union's 5th Framework Programme. Its purpose was to provide a platform to verify and validate concepts for improving fault detection and fault tolerance in process control which emerged from the research. After the project ended, it has been realised that the platform is general and flexible enough to be used in further research on generic asynchronous multitasking systems for distributed control.

How this platform is employed in the design and implementation of a fault-tolerant distributed embedded system is described in Chapter 6. It is shown how certain fault detection and fault-tolerance measures as introduced in Section 1.3.3 are implemented. The distributed homogenous asymmetrical embedded system features full dynamic multitasking with deadline-driven scheduling. To prevent too long latencies in response to events, some critical parts have been implemented in firmware using hardware/software co-design principles.

Finally, in Chapter 7 a consistent implementation of an embedded real-time system, based on a novel and patented operation paradigm, will be described, that was devised and elaborated by a jointly supervised doctoral student at Fernuniversität in Hagen, Dr.-Ing. Martin Skambraks.

5

Hardware Platform

In order to support higher-level studies of control systems' capabilities for fault detection, fault isolation, and fault tolerance, it was required that at the lower level the platform itself exhibited these properties, too. For this reason, first, the existing designs for distributed embedded systems have been studied and proper guidelines set. Then, the architecture has been devised and the prototype implemented.

5.1 Architecture

The architecture of the platform is shown in Figure 5.1. To start with, it was decided to employ the time-triggered CAN protocol (TTCAN [64]; see also Section 3.5.2) as communication infrastructure. To enhance fault tolerance, the communication system was duplicated. In the normal situation, in which the system is expected to be most of the time, the communication load is shared among the two buses. Utilising the fault detection built into the protocol, a communication fault can be recognised. In this case, communication is re-routed to the bus remaining intact. Most likely, because of the degraded performance, it is necessary to reduce the amount of data to be transmitted.

This communication system connects processing units and peripheral interfaces. For implementation, the latter may be grouped into peripheral units; then they require only one (dual) TTCAN bus interface. If they are, however, geographically distributed, or very safety-sensitive, each one may require its own. (There is also possibility of attaching a peripheral interface directly to a processing unit; see S2 in Figure 5.1; this way, however, the fault-tolerance measures given below are not directly applicable.) The processing and peripheral units communicate with each other using the principle of distributed replicated shared memory (Section 3.5.4) as middleware, which provides further fault-tolerance measures.

Apart from these main architectural features, the processing units are connected with further communication means for programming, debugging,

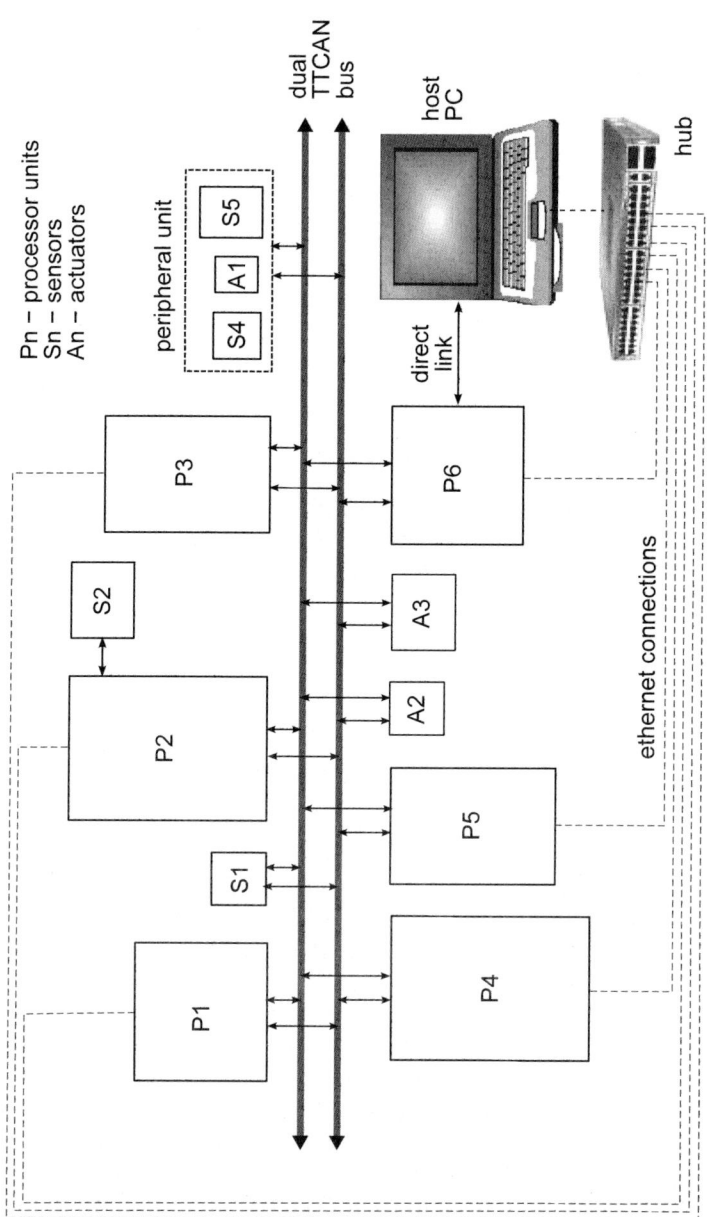

Fig. 5.1. Architecture of the hardware platform

monitoring and diagnostics. For this purpose, a high capacity TCP/IP network is employed. This communication, however, is not fault-tolerant and may not be temporally predictable. In fact, in the final application it is not needed any more. The same holds for the host personal computer that communicates with the distributed embedded platform either by the TCP/IP network, or is connected to one of the nodes, over which it accesses all the others by TTCAN.

5.2 Communication Module Used in Processing and Peripheral Units

To support the TTCAN data transfer protocol, a communication module has been designed and implemented. Technically, it has the form of a piggy-back (or daughter-board) unit plugged into, and attached with screws onto the main-boards of either processing or peripheral modules. The implementation of the communication module and its use in processing and peripheral units are shown in Figures 5.2 and 5.4, respectively.

The physical interface with the two TTCAN buses is implemented with two chips, Bosch TTCAN EC rev. 2.0. The maximum size of a single message as defined by the CAN standard is eight bytes, and the maximum communication speed is limited to 1 Mbps. Figure 5.3 shows a photograph of the interface taken from the side of the key elements FPGA and TTCAN interfaces.

The main part of the communication module, however, is an FPGA chip (Xilinx's Spartan-IIe). Its basic low-level function is to provide an interface between the processing or peripheral units and the communication interfaces connected to the TTCAN communication channels. On start-up, the communication parameters of the TTCAN protocol interfaces are initialised, together with the TTCAN system matrix containing the time-tables of the messages. This is done by sending data, stored in on-board memory, in a stream to the TTCAN interfaces. In case a re-configuration is needed, the communication interfaces are re-initialised by new parameters. This way, the processors are relieved from routine overhead work.

Further, in the FPGA the distributed shared memory is implemented. During operation, the FPGA supplies the contents of the messages for each basic cycle to both interfaces and acquires the received data from them. The memory cells are kept in data storage registers residing in the FPGAs, and are accessible from the processor bus on one side, and the bus interface on the other. When a processor writes a value into a cell, by the firmware in the FPGA it is transparently put into a TTCAN message, which will be transmitted on the network in the next basic cycle. When, on the other hand, a new message is received, the contents are written into the shared memory cells. Not all messages are of interest to all communicating nodes; the FPGA selects the ones to be observed. While writing data into memory cells, certain transformations according to pre-defined functions or operations like time-stamping may

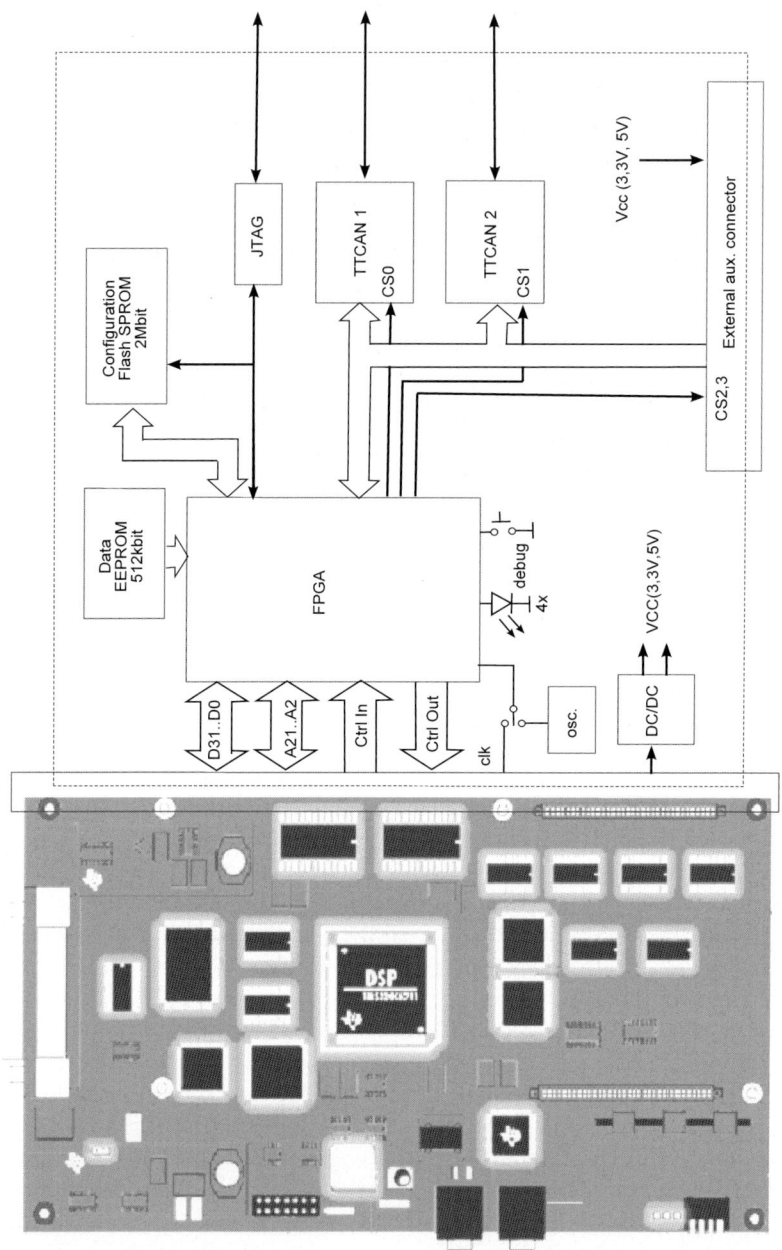

Fig. 5.2. Processing unit with communication module (sketch of the DSP experimental kit by Texas Instruments)

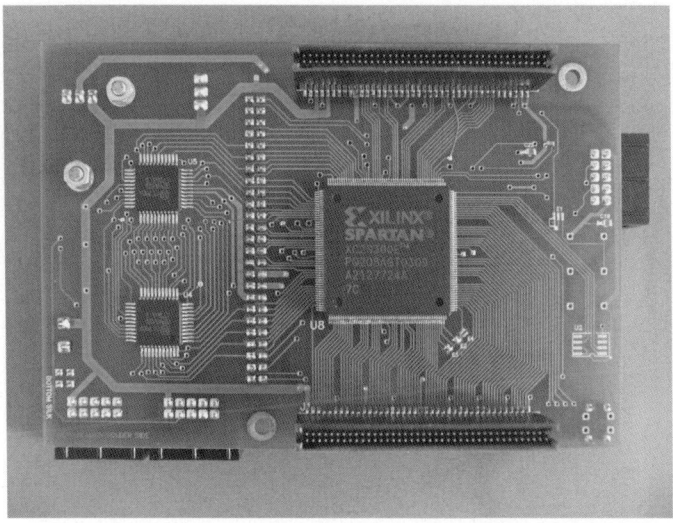

Fig. 5.3. Communication module

be transparently performed within the FPGA. This is particularly interesting in the case of simple peripheral interfaces not capable of such operations.

Beside the processors and the peripheral units, Commercial Off-The-Shelf (COTS) components not having provisions for the specific fault-tolerant communication can be attached to the FPGA and can, thus, be employed in the system. Apart of the FPGA's configuration flash PROM, there is also data memory in form of a 512 kbit EEPROM attached to the FPGA chip. Its purpose is to contain different data which will be used, *e.g.*, for automatic (re-) initialisation of the communication interfaces and other functions to be performed directly within the firmware. The daughter-board also provides an interface to the outside world over a boundary scan connector (JTAG[1]). The processing units have their JTAG ports connected in a daisy chain. During the development and debugging phases, a PC can load configuration data into a module, or it can perform some monitoring functions. To allow for further extensions, an additional external connector was prepared.

Processing units are based on the digital signal processor (DSP) TMS320C6711 of Texas instruments. The experimental board C6711DSK Starter Kit [63] was used, which has the possibility of adding daughter-boards by exhibiting the complete processor bus on two outside connectors where a communication module is plugged in (Figure 5.2). For programming the board in ANSI C, Code Composer Studio V2.1 by Texas Instruments was used.

---

[1] Acronym for Joint Test Action Group; often meaning IEEE 1149.1 standard "Test Access Port and Boundary-Scan Architecture" for test access ports, originally used to test printed circuit boards by boundary scan.

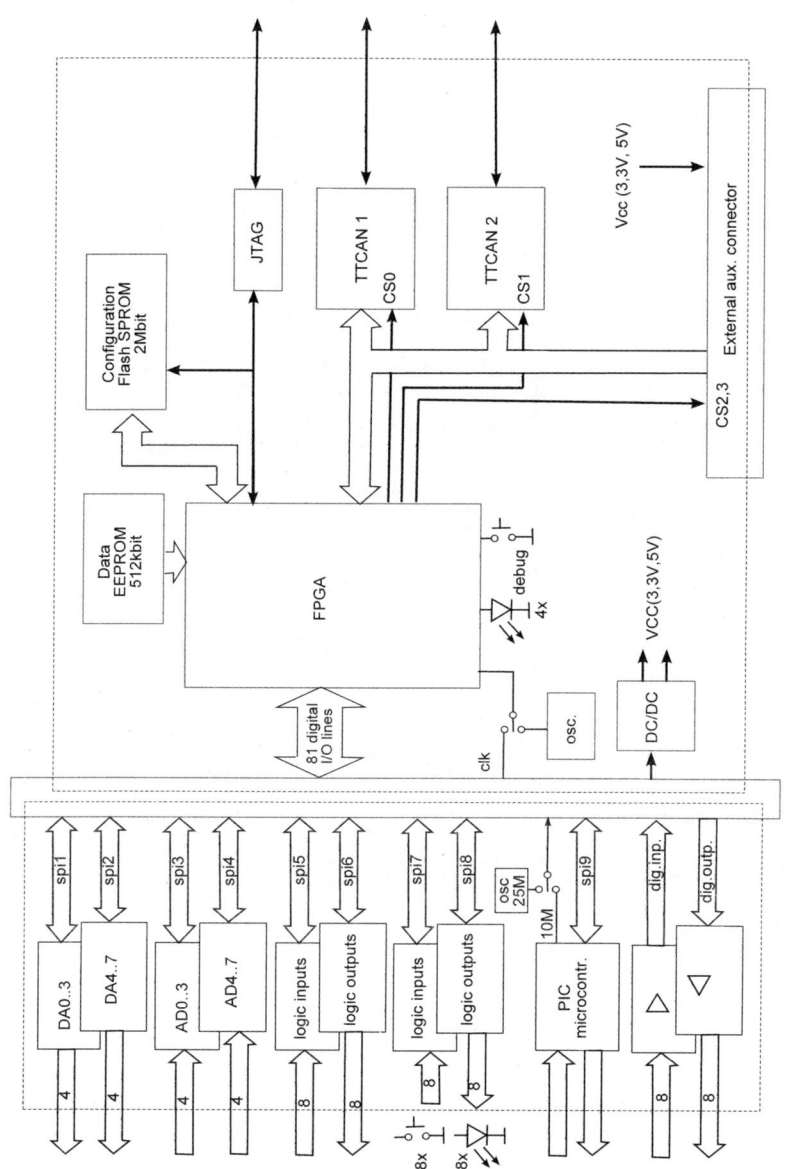

Fig. 5.4. Processing unit with communication module

Similarly, the peripheral units were designed in form of a main-board into which a communication module is plugged in as daughter-board (Figure 5.4). On the experimental peripheral units, a number of analogue-to-digital and digital-to-analogue converters is provided together with pure binary inputs and outputs in different technologies. Such peripheral I/O allows for quick prototyping of process control applications. Apart from that, there is also a simple microcontroller from the PIC®series (Microchip) to emulate intelligent peripheral interfaces.

Fig. 5.5. The prototype

The photograph of the prototype in Figure 5.5 shows three stacked processing units with communication modules, connected by two clearly visible CAN bus cables (the lowest board is detached) with the universal peripheral unit (on the right).

5.3 Fault Tolerance of the Hardware Platform

There are a number of features, introduced by the use of the FPGA, that enhance the system's fault tolerance. It will be shown here how certain principles of fault detection presented in Section 1.3.3 can be employed by the communication module.

In the firmware, some fault detection and isolation functions have been implemented. For instance, data written to a shared memory cell and later broadcast on TTCAN to other replicae, can be checked for certain specified properties and/or behaviour. Plausibility of data values is easily checked by comparing their values with pre-loaded boundaries. Also, dynamics of signal behaviour can be monitored by comparing previous values with new ones; in case of abnormal changes a fault may be suspected. For extremely critical applications, the technique of checking pairs may be used to detect faults. Several (possibly diverse) sensors may acquire process information and send it in separate messages. The FPGA then performs checks whether the data received match sufficiently. Also, to cope with disturbances on the buses, the data redundancy principle can be employed: data may be transmitted in more than one TTCAN message and on both buses, and the conformity is checked by the FPGA.

Further, if there is a probability of communication errors, loop-back testing may be implemented where information is transmitted back and forth, and verified transparently by the firmware. A node may even be designated the function of a bus monitor and verify the general behaviour of data on the buses. Distributed shared memory inherently supports fault tolerance by reconfiguration. If a unit fails, another one may notice the absence of reasonable data in a message and copes with the situation. This is directly implemented in the FPGA. Last but not least, for the complexity of programmed systems, it is difficult to prove their correctness. Since FPGAs are not based on sequential software, and since their functions are simpler and more loosely coupled, they could be verified more rigorously with the usual hardware design and verification means.

5.4 System Software of the Experimental Platform

To support adequately execution of application programs on the hardware platform, appropriate system software has been developed. It resides on each processing unit and is executed in close co-operation of both hardware and firmware as described above. On the peripheral nodes, all required system software functionality is performed by the hardware.

The system software can be divided into several modules or layers: system support layer, communication layer, distributed shared memory layer, and application support layer (Figure 5.6). The system support layer provides an interface to hardware and firmware on one side, and the application software on the other, including the basic operating system kernel functions. The communication layer integrates support for the distributed shared memory mechanism described earlier (Section 3.5.4). The development support module is used for debugging and diagnostics.

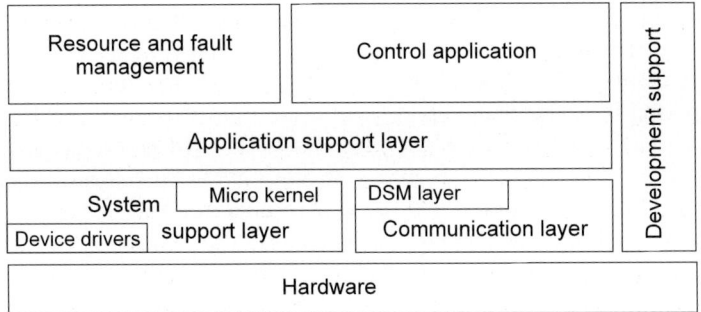

Fig. 5.6. Conceptual organisation of experimental platform's system software

System support layer. This layer provides the basic functionality of the system. It communicates directly with the hardware, and (indirectly) through sensors and actuators with the environment. On the other hand, it is also responsible for smooth execution of applications. It consists of a microkernel, device drivers, and a set of initialisation routines. The microkernel performs basic operations, as can be found in any operating system, however on a considerably reduced scale. Some of these operations are memory allocation, task scheduling, and process synchronisation. In the applications, for which the platform is designed, only simple memory allocation functions are needed. For temporal predictability and simplicity, *e.g.*, support for virtual memory management has not been included. Most of the memory needed is allocated during initialisation.

The early IFATIS version of the system software executes the application code periodically as a single task and synchronously with the communication subsystem, and simulates the tasking operations inside the application. The resource- and fault-management routines are compiled together with an application. Later, most of the microkernel functionality has been migrated to the firmware; see Section 6.3. Proper support for multitasking has been added using the concept of Fault-Tolerant Cells described in Section 6.1. It employs earliest-deadline-first (EDF) scheduling implemented with dedicated hardware components that also perform run-time schedulability analysis of the set of ready tasks in a fraction of the time needed by software routines. Resource monitoring and fault-management has also been put in separate modules and implemented mainly in firmware.

For inter-task synchronisation, the microkernel includes static semaphore objects. In its first version, they have been implemented with low-level BIOS routines that came with the processor board. This kind of synchronisation, however, is limited to the tasks on the same processing board. Later, utilising the shared memory mechanism, inter-module task synchronisation has been implemented. Also, the synchronisation routines of the BIOS have been replaced with more efficient ones.

The device drivers on the peripheral units are pre-programmed, and executed by firmware (*i.e.*, as discrete logic circuits in FPGAs) and/or by simple microcontrollers. The functionality of the microcontrollers implemented by separate chips in the first version was later migrated into the FPGAs. Because of its simplicity, only a few percents of the FPGAs' logical blocks were necessary.

At system start-up, routines are run to initialise the hardware and basic internal data structures, to prepare the communication infrastructure, *etc.* As a part of initialisation, a set of self-diagnostic routines is executed to detect possible failures in the hardware. System initialisation is based on the configuration tables provided by the application development tool. The tables include general information about each node (*e.g.*, ID, type and priority), information about integrated peripherals, principal attributes for the communication infrastructure, and so on.

Communication layer. Communication functions perform all communication between modules of the system. The timetables for the TTCAN protocol are provided in advance by the development tool. A timetable consists of a sequence of slots used to transmit individual messages. A message represents a piece of information that is transferred between different processing nodes, sensors and actuators. The communication routines write to and read from slots. In the first version of the system software, a small amount of communication bandwidth was set aside for debugging messages. Later, TCP/IP communication links were used for this purpose.

Distributed shared memory layer. Tightly integrated with the communication layer is the subsystem that implements the distributed shared memory model. It serves as middleware, and allows for independent virtual communication links between source and destination of any message. When both peers reside on the same node, communication is performed as memory-to-memory transfer. When intra-module communication is required, the TTCAN communication protocol is used.

Application support layer. This layer allows control applications to interact with the hardware by a set of pre-defined API routines. System interface routines are used by applications to invoke services of the system software. Generally, all functionality of the underlying system software can be used by applications if needed. However, it is expected that most of these routines will mainly be used by the resource and fault-tolerance management modules. The routines are integrated into the object library and are linked to the control applications. The interface to the system routines is provided by means of C header files.

Support for development, testing and diagnostics. The hardware platform provides an experimental test-bed for different higher-level approaches to deal with faults in control systems. To this end, adequate system support for testing and diagnostics has been implemented. It can be used for monitoring and setting values of shared memory cells and certain global variables in control applications. The changing of information can be monitored and recorded during a selected period of time. Then, data can be downloaded and analysed off-line. In the early hardware implementation, the debugging routines on a dedicated processing node communicated with the development platform through a JTAG port. If required, the commands were relayed to other nodes in the system by means of the communication subsystem.

On each node a simple software routine accepts diagnostic messages, performs the operations requested, and returns information to the requesting nodes. A message represents a single command, containing command code, recipient node identification, and eight bits of data. The development and diagnostics support system implements three commands to read memory locations in processing nodes (GET_LONG, GET_WORD, GET_BYTE), and three commands to write into such memory locations (PUT_LONG, PUT_WORD, PUT_BYTE). Each of them includes local memory address, and data in case of write operations. As reply, the same message is sent back by the receiving node, with the new data in case of read operations. These commands can be used for basic memory read/write operations, loading of programs, *etc.* Since the messages require absolute addresses, an additional command (GET_STRUCT) is used by the development system to acquire the start addresses of all mandatory data structures on the nodes. Beside the messages mentioned there are several for off-line data acquisition and monitoring. Currently, there are 24 different system commands implemented on the processing nodes and used by the development platform. A set of commands is reserved for user applications. These messages are processed independently of system software by a procedure written in C code.

The overhead for diagnostics should not interfere with normal system operation. In particular, the transmission of the debugging messages should be performed independently from normal inter-node communication. Therefore, all diagnostic messages are considered as sporadic messages with low importance, and are transmitted in the arbitration message slots described in Section 3.5.2.

The development support and diagnostics module was used during development to load application code from the development system onto the processing nodes. For this purpose, a simple loader was implemented. The compiler creates object files in common object file format (COFF). The object files are then read by the loader, prepared for execution, and put into memory. When an application has become stable, the code is pre-loaded into the EEPROMs.

The TTCAN communication protocol is not suitable for fast transfer of large amounts of data. The maximum size of a CAN message is eight bytes, with the transmission speed limited to 1 Mbps. Although some tricks were utilised (*e.g.*, an unused part of the message ID was employed to carry the command code), the loading of program code and debugging were slow. Consequently, an alternative solution has been implemented. A dedicated Ethernet hardware communication module has been added to each node, which allows TCP/IP communications with the development platform. In control applications, these modules cannot be used for process data transfer for several reasons, the most important being that the basic TCP/IP communication protocol lacks temporal predictability.

6

Implementation of a Fault-tolerant Distributed Embedded System

In this chapter it is shown how the platform has been employed in the implementation of an intelligent re-configurable fault-tolerant control system within the research project [62].

The project's major goal was to establish a framework of intelligent fault-tolerant control technology, providing engineers and scientists with advanced methods to address fault-tolerance problems in complex control systems in a systematic way. It was intended to build a basis for the development of user-specified tools and systems in all kinds of industrial sectors. The project's second goal was to develop a novel methodology and a software tool for integrated design of real-time, multilevel fault-tolerant control systems. The final goal of practical prototype implementation is to customise the results results for automotive systems, a water-treatment process, and for constructing a fault-tolerant control platform for robotic systems.

Besides enhancing the quality and robustness of process components, using hardware redundancy is a traditional way to improve process reliability and availability, which has been extended to the use of software redundancy during the last decades. Due to the fixed structure and high demands for hardware and software resources, the applicability of redundancy strategies is usually limited to some special technical processes, or to parts of key system components like central computing unit or bus system. Evidently more flexible and effective are the fault-tolerance strategies with fault accommodation or system and/or controller reconfiguration. In a fault-tolerant control system, early fault detection and isolation plays a key role.

The objectives of a part of the project which was entitled "Fault Detection and Isolation – Fault Tolerance in Computer Systems" were (1) to analyse the standards and guidelines, and the needs of control engineers in order to detect and isolate faults from the point of view of computer control systems, (2) to develop software modules aiming at designing hardware and software architectures appropriate for fault detection and isolation, and providing high system integrity employing fault tolerance, and (3) to implement a prototype computer control system platform to evaluate and validate these principles. The resulting hardware platform has been described in detail in Chapter 5.

After conclusion of the project, further generalisations have been made in order to make the scheme for fault detection and tolerance universal and more flexible. Below, the resulting model and an implementation for distributed control systems are described. They are based on the existing hardware platform and on the guidelines that emerged from the project.

6.1 Generalised Model of Fault-tolerant Real-time Control Systems

In the control community, the most widely used technique to describe a control system is based on logical function blocks. This approach is also supported by most generic and custom design tools for control applications. Each function block can have several inputs, performs some function, and produces a number of outputs. Several function blocks can be combined in different topologies, each function block can be decomposed into more primitive blocks, *etc.* This method is used both for theoretical system analysis and to design and implement control applications. Moreover, this model-based approach becomes more and more popular for dealing with all kinds of computer systems. There are further benefits of using function blocks. When designing distributed systems, the innermost function blocks represent basic units of allocation to processing elements. A composite function block, on the other hand, can be executed in distributed fashion. Furthermore, on the top (system or context) level an entire control system — distributed or not — can be considered as a single function block.

A generalised model of a control system needs to take fault tolerance and real-time issues into account that are typically not integrated in the traditional development methods for control applications. The model to be presented has been designed for easy implementability on diverse hardware and for matching well general distributed architectures. It supports hardware/software co-design, and allows for a smooth transition from logical to physical architecture. Its basic idea is to combine control functions (composed of function blocks) with a monitoring function and reconfiguration capabilities, and to include the notion of system resources. The monitoring function supervises the behaviour of input and output signals, and intermediate system states. As a result, the so-called Fault-Tolerant Cell (FTC) is introduced; see Figure 6.1. An FTC combines both the algorithmic part of a control application and all resources needed to perform the task given. The algorithmic part is represented by a set of function blocks. In addition, an FTC also encloses the fault tolerance and temporal features of the encapsulated blocks.

On the context level, an entire control system is considered as a single global FTC that contains all hardware resources and program code. Its objectives are expressed generally (*e.g.*, "control a chemical process"). This FTC

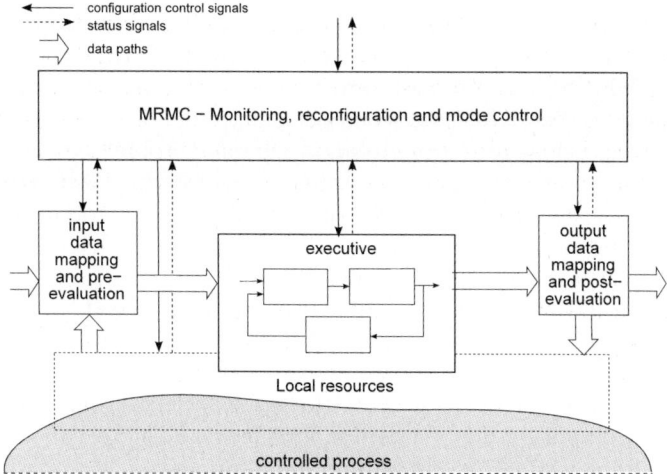

Fig. 6.1. Structure of a fault-tolerant cell

can be decomposed into a set of more primitive FTCs that perform a number of subtasks of the primary objective. Instead of running the control application code, the global FTC is managing its subblocks. Each subblock owns a subset of resources and a part of program code allocated to the superior block. On this level, the objectives are more specific (*e.g.*, "maintain the temperature of a mixture"). This decomposition can be applied recursively, *i.e.*, FTCs can be decomposed hierarchically. On the lowest level of the hierarchy, application code of control functions is executed. Since a specific objective can be accomplished in different ways, it is not necessary that all subblocks are active at a specific point in time. Selecting alternative function blocks and/or alternative FTCs also provides for both fault tolerance and timeliness.

The internal configuration of an FTC at a specific instant is selected by configuration control signals provided by higher levels in the FTC hierarchy, or by the operator on the top-most level. Each configuration also inherently defines certain parameters about how the given objective or task should be performed (*e.g.*, execution deadline, or accuracy of results). In the case of hard real-time tasks these parameters are firm. For soft real-time tasks, some of them can be given in ranges, or by some fuzzy rules. This allows for better flexibility in cases that only limited resources are available or of transient overloads. If there are not enough local resources available to perform a selected configuration, or if the required execution parameters cannot be met, appropriate status information is generated to the upper levels in the FTC hierarchy and, eventually, to the operator. Consequently, at an upper level, alternative scenarios can be chosen, the general objective can be degraded, or the system can be shut down in a controlled manner. If an FTC can handle hardware, software, and timing faults locally, a higher level of hierarchy is not notified.

Different topologies of FTC interconnections can be implemented (*e.g.*, centralised control, master-slave or various multilayer organisations). On hierarchically different levels, usually control configuration and status information is exchanged between FTCs, only. On the same level, FTCs mostly communicate by passing data from outputs to inputs of other FTCs.

The internal structure of FTCs is shown in Figure 6.1 In general, any FTC acquires some input information, processes it, and produces outputs. The inputs are read either from sensors in local resources, from operators, or from other FTCs. Similarly, the outputs drive actuators or constitute input data for other FTCs. Traditionally, sensors, actuators, and other resources are not considered to be parts of function blocks, as is the case for FTCs. For instance, an FTC measures temperature instead of just reading a sensor. Since the corresponding sensor is part of the FTC's local resources, in case of a fault the temperature may be determined in an alternative way, if possible without the need to notify the superior FTC. A conversion function (*e.g.*, from an acquired raw integer to a physically correct floating-point temperature value) may be defined within the FTC, together with bounds and/or other plausibility control. For an alternative sensor, a different function may apply. Similarly, FTCs influence the plant through changing some physical properties (*e.g.*, by heating the liquid in a tank, or by increasing/decreasing fuel flow). This approach allows for a higher degree of flexibility and for better fault tolerance.

A Fault-Tolerant Cell is composed of:

Local resources. Any FTC is associated with a set of resources: processing capabilities, input/output interfaces, sensors and actuators, *etc.* Some of the resources are available exclusively to a single FTC, and some can be shared among several FTCs. For example, more than one FTC can be executed on a single processing element, or sensor data can be read by several FTCs. An actuator, however, will most likely be associated with one single FTC, only. A set of resources available can change during local or global reconfiguration of a system. The resources allocated to a specific FTC are available in all its subblocks as well. There can be no resources locally assigned to a subblock, which are not part its superior block, too. Therefore, the FTC on the context level owns all resources of the control system.

Monitoring, reconfiguration and mode control (MRMC). This module manages the FTC. Based on external configuration control signals, the MRMC allocates and configures available resources and other components of the FTC to perform a requested task. According to the configuration selected, the programs performing the control algorithm are invoked. The MRMC also monitors all associated resources and traces any malfunctions in them. If a fault is detected by MRMC during the execution of the FTC's executive, first, the MRMC tries to deal with it locally by using redundant or alternative components, by reconfiguring the local subblocks, *etc.* If a fault cannot be dealt with locally, the MRMC immediately notifies the superior entity by means of appropriate status information.

Input data mapping and pre-evaluation. This module receives all input data signals needed by the FTC to perform a given task. The inputs originate either from the local resources or from external information paths. It maps external data values into the data domain used by the FTC. For example, it can map integer values produced by an A/D converter into a floating-point quantity that represents the actual control parameter. The module can check the plausibility of input data and inform the MRMC if anomalies are detected. It also serves as a multiplexer if alternative information sources for some input exist. For instance, several redundant sensors may be allocated to the FTC. After start-up, one of them serves as actual data input. If this sensor fails, the MRMC can transparently switch to the next available one without the need to restart the primary control function. In some cases, if no alternative for some input is available, the MRMC can activate an auxiliary module that will provide the missing information by an approximation based on other sensorial inputs. This information is then delivered to the same data input as if it came from the sensor. Not all input data needs mapping and probably not every signal is tested.

Output data mapping and post-evaluation. This module takes the values produced by the cell and converts them into a form appropriate for use by other FTCs or local output devices. A floating-point quantity, for instance, that represents opening of a throttle in percent must be converted into an appropriate integer value to be used by the D/A converter in the throttle actuator. Again, output signals can be checked for any anomalies and the MRMC can be notified. Similar to the input data mapping module, this one may also utilise redundant actuator elements.

Executive. The executive performs the FTC control functions according to the configuration parameters set by the MRMC. On non-leaf levels of the block hierarchy, the executive consists of several subordinate FTCs. Here, the MRMC controls which subordinate FTCs should be activated, and how they interact with each other. On the elementary level, the executive executes program code of control functions on the processing element, some dedicated hardware component, *etc.*, which are part of the local resources. In either case, the executive also provides some information to the MRMC if required. As noted before, sometimes the MRMC should know if subordinate FTCs cannot execute their tasks within required deadlines, or if subordinate FTCs do not have the capability to perform the tasks at all.

6.2 Implementation of Logical Structures on the Hardware Platform

As previously discussed, a control system as a whole can be portrayed by a single global FTC. All resources and control functions are allocated to this FTC, and its MRMC monitors and manages the control application. However, it is not likely that a complex application is executed as a single executable program. Therefore, the executive of the global FTC is decomposed into several composite (*e.g.*, FTC1 in Figure 6.2) or elementary FTCs. Both can be mapped onto one or several processing units, and one or more peripheral devices can be associated with them (*e.g.*, FTC3 and FTC7). It is also possible that several FTCs share a single processing unit (*e.g.*, FTC5 and FTC6). This is particularly convenient in a case of tightly coupled partial processes where communication delays may introduce jitter.

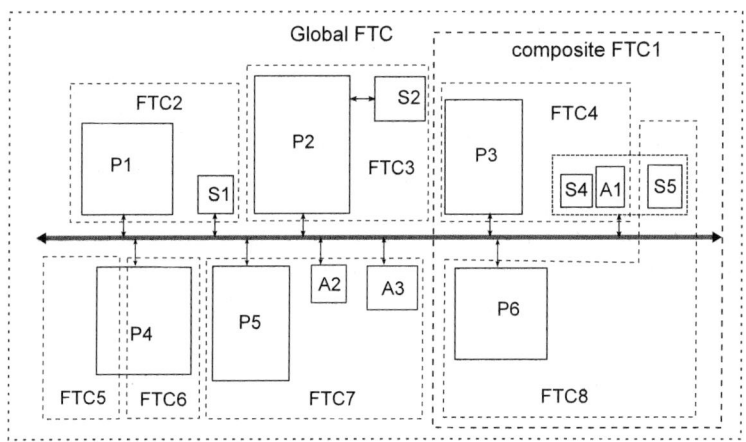

Fig. 6.2. An example of mapping different FTCs onto the distributed hardware architecture

To provide for the necessary system integrity, a mapping is not permanent and can be reconfigured in the case of faults. Control functions, as well as monitoring and other blocks of FTCs must be executed on the resources that are proven faultless. Reconfiguration is managed by higher-level MRMCs based on the status information received over the communication network from the subordinate FTCs. Alternative or redundant resources are employed, if available, or the system's performance is degraded gracefully in order to survive the situation. However, the set of all possible system configurations is static and known in advance. This set is prepared and analysed for temporal properties off-line during the development phase concerning potential faults and considering required deadlines. Of course, this can represent a very complex task, and should be supported by appropriate development tools. Particular attention must be paid to the implementation of the global and local MRMCs, which is not explicitly shown in this scheme.

Global MRMC (G-MRMC) monitors and controls the fault-tolerance behaviour of a control system as a whole, based on the status information provided by other MRMCs and the operator. Possible implementations of G-MRMC are:

Centralised G-MRMC. The basic implementation is a single centralised G-MRMC. Status and control information is transferred to and from the G-MRMC over the communication infrastructure. A single G-MRMC also represents, however, a central point of failure, requiring a fault-tolerant implementation: when the node executing the G-MRMC fails, its functions must be passed to any other processing unit. To allow for that, several nodes execute the same G-MRMC code in parallel, but only one replica generates the system's reconfiguration control information.

Distributed G-MRMC. If, as in the present case, the communication protocol is based on the message broadcasting mechanism, this is a better solution to implement the G-MRMC. All messages sent from a node can be received by all other nodes in the system. Thus, the status information messages can be processed and reacted upon by any node locally. Therefore, there is no need to transmit any control information for reconfiguration. Moreover, only the status information relevant for a specific processing unit must be processed. This approach may simplify implementation, but has the drawback of increased workload on every processing node.

Lower-level MRMCs monitor and control the fault-tolerance behaviour of their executives. Since it is possible to allocate a single FTC to several processing units, similar considerations related to fault tolerance arise as for the G-MRMC. However, the same solutions apply, and the distributed MRMC can be implemented on all associated nodes.

Based on the current operation mode, the MRMCs of the elementary FTCs active on a particular processing unit also determine the set of executives that must be executed on this processing node, and its timing constraints. Similarly, these MRMCs also reconfigure other resources of the node (*e.g.*, locally connected sensors and actuators, communication data paths, or pre- and post-evaluations).

6.3 Partial Implementation in Firmware

Executing MRMCs, microkernel, communication subsystems, *etc.*, takes a certain amount of a processing unit's capacity away from the application programs. Further, the complexity of the temporal analysis is increased and, in the case of improper design, temporal predictability may be jeopardised. This overhead can be reduced if additional processing facilities in form of dedicated hardware are employed, whose functions are defined by corresponding firmware. It resides at the system bus of the processing and peripheral nodes to allow them to access its dual-port memory cells. Network interfaces are connected as well. This hardware provides some low-level functionality, enhanced fault tolerance, modularity, and increased performance, and implements fault-tolerant inter-module communication using the approach of distributed replicated shared memory. It can also perform parts of control applications (thus supporting software-hardware co-design), and serve as a gateway to Commercial Off-The-Shelf devices, which do not comply with the fault-tolerance requirements. For its realisation, a FPGA device has been employed, which implements a number of hardware modules on a single chip, running in parallel and independently of each other. The outline is shown in Figure 6.3 (*cf.* also Figures 5.2-5.4). In the following subsections, implementations of three main modules will be presented.

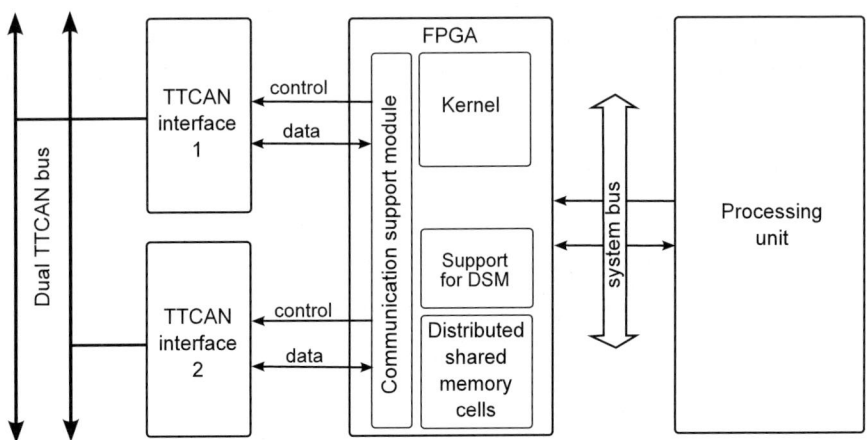

Fig. 6.3. Firmware-implemented communication support module

6.3.1 Communication Support Module

A time-triggered bus requires the periodical attention of a controller; if it were run by a processing unit, it would influence the unit's performance (although the delay would probably be predictable). Instead, this module controls all communication hardware interfaces. It handles buffering of input and output data, and monitors proper operation of the communication buses. Due to the relative complexity of the communication chips used in the platform, a solution with digital circuits was not possible; instead a simple microprocessor was implemented in firmware. It manages TTCAN interface chips, and transports data between them and the registers of the distributed shared memory; see Figure 6.3. This is carried out transparently and in parallel with program execution in the main processor.

The microprocessor mimics the architecture of the 8-bit Microchip PIC® microcontrollers [83] and uses the same instruction set. This enables one to use any existing compiler for this family of microcontrollers. It consumes a very small amount of FPGA resources – theoretically it would be possible to implement several microprocessors of this kind in a single FPGA device. In addition to its primary role of supporting inter-module communication, the microprocessor performs some other tasks not possible by simple hardware logic.

6.3.2 Supporting Middleware for Distributed Shared Memory

This module extends the communication support and handles the access to the shared memory cells on one hand and their replication over the communication module on the other. It implements simple data mapping and evaluation, and detects when data in the memory cells are changed potentially triggering some tasking operations.

To detect if a value is in its valid range, logic comparators are used. The logical scheme for this is given in Figure 6.4.

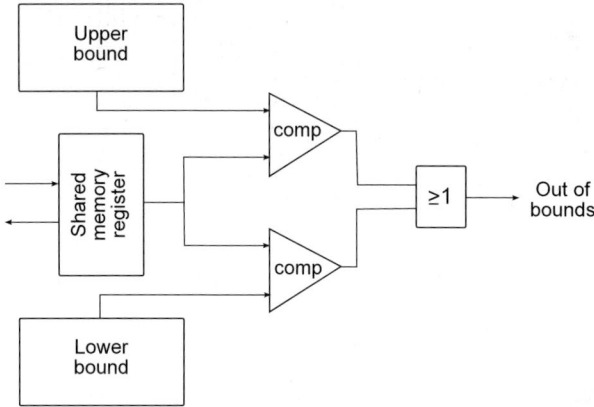

Fig. 6.4. Range checking of transmitted values

For integer quantities the implementation of the evaluation logic is straightforward. Consumption of FPGA resources can be reduced if only a coarse comparison is performed. In this case, several most significant bits of the data are taken into account. Another simplification is possible if only non-negative values are allowed, and data is only limited by an upper bound. Further, some resources can be saved if pre-defined constants are used instead of registers to represent a lower and/or upper bound. However, the reduction achievable by this approach is minimal and reduces the flexibility to use the circuit for different cases.

On the other hand, for floating-point values, the extensive amount of silicon needed for implementation may be prohibitive. The comparison of two floating-point values requires equalisation of their exponents, subtraction, and normalisation of the results. This is much more effectively performed by the arithmetic unit of the main processor. Therefore, evaluation tests should be performed by the software. There is an exception to this rule if only the relative magnitude of floating-point values is observed. In this case, the exponents of the numbers can be evaluated independently of their mantissas, which can be implemented with basic integer logic.

One of the ideas that have been tested in this context is automatic generation of evaluation logic in FPGAs. Based on the data types and magnitudes of data provided by specifications, the appropriate VHDL code should be generated automatically where possible.

6.3.3 Kernel Processor

To implement (partially) the kernel processor's functionality, a dedicated processing unit specialised to provide operating system functions has been implemented in firmware. By this co-processor approach, maximum benefits can be achieved, particularly in the area of distributed systems [81].

The most time-consuming function of the microkernel is to establish which task must be executed next, considering remaining execution times and deadlines. With dedicated hardware this can be executed in only a fraction of the time that would otherwise be needed by the main processor. However, due to limited memory available in the FPGA device, only the most essential information about tasks can be stored inside it. The full contexts of the tasks must be kept in the main processor's memory. The same hardware logic can also detect any deadline violation. Although the EDF scheduling algorithm has been employed, any feasible scheduling mechanism can be implemented.

Determination of the next task to be executed. According to the EDF scheduling policy, the task with the nearest deadline must be executed first. The most simple method to determine this is by sorting the tasks by their deadlines. Since deadlines are fixed, there is no need to check the conditions continuously. The situation changes only when a new task becomes active or when a task terminates. This is in contrast to some other scheduling policies where task priorities are changing constantly (*e.g.*, least-laxity-first). As tasks may change their states within specific system routines, re-scheduling can be integrated into them. The hardware implementation is a matter of realising a minimum function that compares deadlines of ready tasks and finds the smallest (nearest) one. This can be presented by the algorithm shown in Figure 6.5.

Should the set of ready tasks be empty ($n = 0$), the resulting task's index is 0, meaning that no task is scheduled.

This code can easily be translated into VHDL. Since integer arithmetic is preferable for common VHDL solutions, the deadlines should be represented as integer values (*i.e.*, as numbers of milliseconds from a specific point in time; for details *cf.* Page 114).

Scheduling of the ready tasks and feasibility check. For real-time systems it is very important to determine whether the current set of active tasks is

```
min_task_index = 0
min_deadline = ∞
for i=1 to n do
    if taskinfo[i].deadline < min_deadline then
        min_deadline = taskinfo[i].deadline
        min_task_index = i
    end if
end for
```

Fig. 6.5. Sorting ready tasks

schedulable, *i.e.*, if all its members will meet their deadlines. For EDF this is determined by Equation 2.3 on Page 42. This equation, however, contains relative response times. On the other hand, the deadline of a task, from which its response time may be derived, is set to some absolute time instant when the task becomes ready. To simplify implementation, Equation 2.3 is converted into a form employing absolute deadlines by adding the current time t:

$$z_k \geq t + \sum_{i=1}^{k} l_i, k = 1, .., n \tag{6.1}$$

This equation states that the sum of the remaining execution times l_i of all tasks T_i scheduled to run before, and including T_k, added to the current time, must be less or equal to the absolute time of deadline z_k of task T_k (the cumulative workload to be performed prior and during execution of task T_k must be completed before its deadline). The tasks are sorted by their ascending deadlines. Again, the condition determined by the formula is static, and must be re-evaluated only when one or more of the tasks change their states.

The schedulability check can be converted into a form with a double nested loop, *cf.* Figure 6.6

The innermost loop is used to determine the (next) task with the minimum deadline. This determines the order of task execution. Then feasibility is tested by first adding the remaining execution time of the current task to the cumulative execution time and comparing it to the task's deadline. With slight adaptation this algorithm can easily be translated into VHDL code. The drawback of this approach is its speed. Implemented with VHDL, for N tasks the execution requires $N^2/2$ iterations of the innermost loop, and each iteration requires several clock cycles.

Another solution more appropriate for hardware implementation is to perform all operations of the algorithm's loop in parallel with pure hardware logic. The idea is to keep the ordered list of tasks in a shiftable hardware queue, sorted by their deadlines. The queue elements queue can synchronously be shifted to the right, and the parameters of all queue items can synchronously be compared with a value. In each step one task from the unordered set of ready tasks is considered: its deadline is simultaneously compared with the

```
cumulative_finish_time = current_time
for i=1 to n do
    for j=i+1 to n do
        if taskinfo[j].deadline < taskinfo[i].deadline then
            swap(taskinfo[i],taskinfo[j])
        end if
    end for
    cumulative_finish_time = cumulative_finish_time
                             +taskinfo[i].remaining_exec_time
    if cumulative_finish_time > taskinfo[i].deadline then
        raise deadline violation error
    end if
end for
```

Fig. 6.6. Schedulability check

ones of the tasks already in the queue. This way, its position in the list is determined. Then, the tasks with later deadlines are, like in a shift register, shifted one place, and the parameters of the scheduled task are stored in the gap emptied. To test feasible schedulability, additional synchronous logic is implemented to maintain cumulative execution time.

Since tasks are sorted in ascending order of their deadlines, the remaining execution time of the task T_k influences only the schedulability conditions of itself and the tasks behind it. Therefore, it must only be considered in the feasibility check of the tasks kept in queue elements that are shifted in the course of storing task T_k. For the check, the remaining execution time of T_k must be added to the cumulative execution times of all subsequent elements, which can be done in the third execution cycle. To convert relative execution times into the form of absolute time, the current absolute time at the instant of scheduling is added to cumulative relative execution time of the first task in the list when this cell is loaded. After summation, the cumulative times are compared with the deadlines of the queue elements; those items are marked which will cause deadline violations. To summarise, the processing steps mentioned constitute the following algorithm:

Step 0: Invalidate all elements in the list. This step is carried out only once. Then, for each active task execute the next four steps.

Step 1: Compare the deadlines of all elements already in the ordered list with the current one, and mark elements whose deadline is greater than the current one.

Step 2: Shift all marked elements one place towards the end of the list, and fill the gap with the values of the current task. Update (set) the mark of the last element in the list to be included in the next step. Set the cumulative execution time of the element being updated to the cumulative execution time from the previous cell. If the first element in the list is updated, set the cumulative execution time to the current time of the system instead.

Step 3: Add the remaining execution time of the current task to all marked elements.

Step 4: Compare the cumulative execution times with the corresponding deadlines and find elements whose deadlines will be violated.

The logical scheme of a single list element is shown in the Figure 6.7.

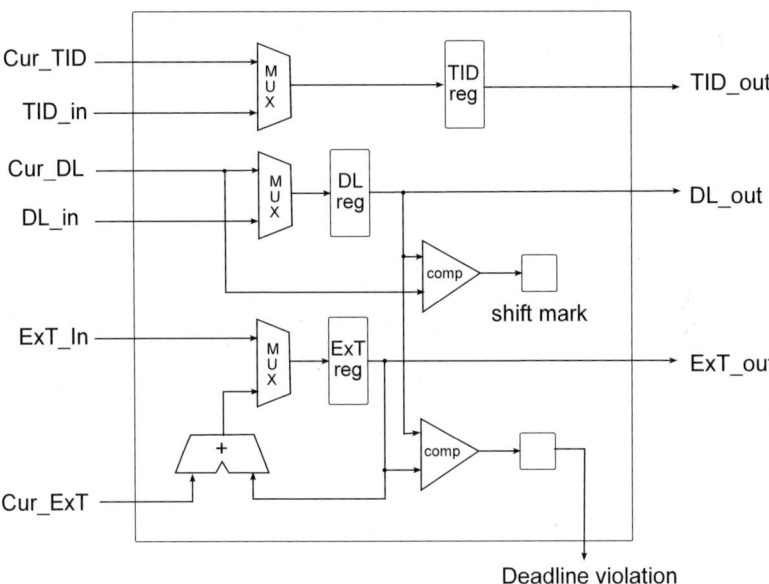

Fig. 6.7. Element in the list representing a task schedule

Each evaluation element contains registers to keep a task's ID, its deadline, and cumulative execution time, and Cur_TID, Cur_DL, Cur_ExT represent task ID, deadline, and remaining execution time of the task currently being evaluated. In the first step of the algorithm, the upper comparator in Figure 6.7 compares the value of the deadline register with the deadline of the evaluated task and sets the shift mark flag, if the former is lower. The inputs and outputs denoted by the _in and _out suffixes are used to transport (shift) values from one element to the subsequent one during the second step. The multiplexers MUX determine which values will be stored into the element's registers. They can either be inserted from the task currently evaluated, or shifted from the outputs of the previous element. If the shift mark is not set, the values of the registers remain unchanged. The summation unit is used in Step 3 to calculate new cumulative execution times. The deadline violation signal is set in Step 4 if the cumulative execution time register in the element exceeds the value in the deadline register (*i.e.*, the condition of Equation 6.1 is not met).

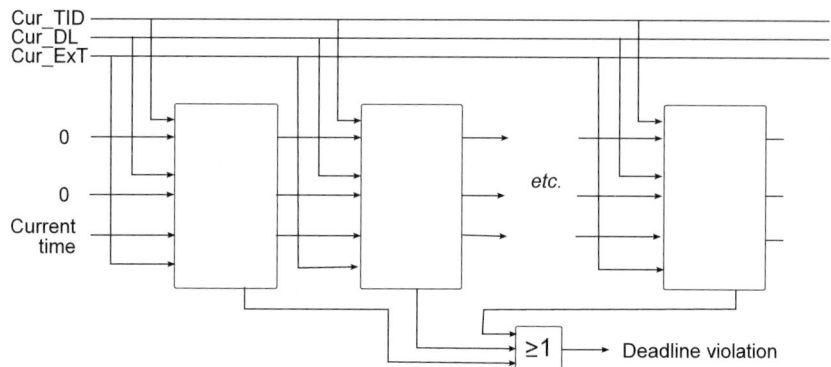

Fig. 6.8. Queue of the task list elements

By coupling several elements in sequence, a list of arbitrary length can be constructed, as shown in Figure 6.8. In the first element, the value 0 is shifted in as an ID of a non-existing task, and also as its initial deadline. Further, the current time is added to the execution times of tasks. Deadline violation signals from all cells are disjunctively combined for common error diagnostics.

According to all this, the algorithm's total execution time is $4 \times N$ clock cycles plus one for the preparation of the data structure. The drawback of this approach is to require a lot of hardware resources (*i.e.*, each element needs two comparators, one adder, and several registers).

Early detection of potential deadline violations. Equation 6.1 for feasible EDF schedulability is valid only if the estimated execution times are accurate. For some reasons it is, however, possible that a task's actual execution time is larger than estimated, which would jeopardise the schedulability of all subsequent tasks. To prevent this, a watchdog timer can be implemented that monitors and limits the actual execution times of tasks. Any time a task is run or continued, the watchdog timer is set to its (remaining) execution time. If the watchdog timer runs out before the task is finished, an error signal is generated.

Event generation. Activation, continuation and other tasking operations are associated with a variety of conditions depending on time, internal or external signals, or combinations of these, see *e.g.*, Figure 4.6. When a condition is fulfilled, the appropriate event is generated, and the current set of tasks must be re-scheduled. The generation of such events can relatively easily be implemented with FPGA devices.

Time events may be generated only once or periodically, during a certain interval of time or perpetually. In addition, the activation of a task may be delayed for a specified amount of time, or until some external event occurs. Thus, the most obvious devices to support timed activations are timers, which are based on counting clock ticks. The logic for full support of time event generation may be implemented as presented in Figure 6.9.

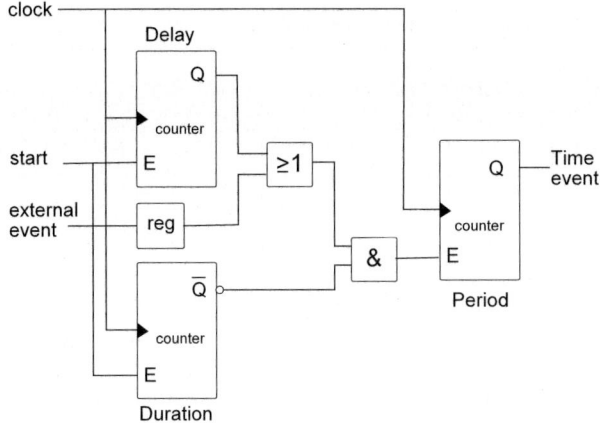

Fig. 6.9. Event generation logic

The Period timer periodically generates time events by loading a pre-set value and decrementing it at every tick of the clock. It is enabled according to the values of the Delay and Duration timers. The Delay timer starts counting, the Duration timer stops it after a pre-determined amount of time. Delay and Duration times are executed only once (*i.e.*, they are pre-set with specified values and stop counting when they expire). Instead of the Delay timer, an external event signal may be used. For aperiodic timing events, the Duration timer is pre-set in such a way that the Period timer triggers only once.

Events may also be generated upon arrival of new data, upon changes of data, upon data reaching specific values, *etc.* To implement this, triggering logic is integrated into data evaluation logic.

6.3.4 Implementation of Monitoring, Reconfiguration and Mode Control Unit

By implementing the global and other Monitoring, reconfiguration and mode control units (MRMCs) in firmware, their overhead is eliminated. Most of the MRMCs' functionality is rather primitive, so that it can be expressed with Boolean functions easily implementable with simple logic, which enables or disables different parts of the firmware.

Similarly, for the parts of system software that cannot be implemented in firmware, appropriate status information is generated. For more complex Boolean functions, the FPGA's dedicated memory blocks can be utilised as look-up tables. In this case, the address lines of the memory are used as inputs and data lines as outputs. The contents of the memory correspond to the truth-table of the Boolean function being implemented. Memory blocks can usually be configured with different widths for addresses and data to accommodate for different numbers of input and output variables. Due to resource constraints, however, only a limited number of MRMCs can be implemented.

6.4 Programming of the FTCs

The MATLAB®/Simulink®suite is one of the most widely used software packages in academia and industry to model dynamic systems. It provides broad support for design, test and simulation of control applications. There are, however, several limitations for building distributed control applications. To cope with them and to provide support for experimenting with different fault-tolerance methods, the toolset has been expanded. One of the features offered by MATLAB®/Simulink®is automatic generation of high-level language program code from Simulink®models, that can be customised for virtually any hardware platform. This is performed by the Real-Time Workshop®package. The generated standard ANSI C code is customised to accommodate specific compilers, input and output devices, memory models, communication modes, and other characteristics that applications may require. It is then compiled using ordinary compilers to produce runable code to be finally loaded onto a platform. It is also possible to directly connect Simulink®models and code executing on a target platform. In the so-called external mode, Real-Time Workshop®establishes a communications link between a model running in Simulink®and code executing on the target system.

6.4.1 Extensions to MATLAB®/Simulink®Function Block Library

To meet the model-based approach used in the research project [62], a specific programming tool based on TNI's ControlBuild®[110] was developed. Apart from model-based programming of control applications for the project's hardware platform, the goal was to support the middleware concept of shared memory cells. To this end, some new function blocks were introduced into the Simulink®library. For these blocks, both code for simulation and for execution on the target was provided.

Node info block. On this distributed platform, application programs are statically mapped onto different nodes in the network. For each configuration, mapping is pre-defined by the designer. Thus, the application model must be divided into subsystems corresponding to specific, possibly diverse processing nodes that have certain properties. To describe their hardware characteristics (*e.g.*, ID, node number, priority, node being master or slave in communication) the node info block has been introduced. The information from this block is integrated with the code generated for the corresponding subsystem.

System status block. The information about the current execution mode of the system is provided in yet another block. It can be used by applications to determine the current system state, and to switch on and off different application subsystems by means of different enable signals. This block has only a single output, whose actual value is provided by the system software. No additional parameters are required.

Working with shared memory. To implement the shared memory model described above, two additional function blocks were added to the Simulink® library: one to read from and one to write into a specific memory cell which, in MATLAB®/Simulink®, is defined as a variable. The custom dialog window is used to define specific parameters: type of data to be sent or received, shared memory slot identifier, fault-tolerance options (*e.g.*, whether the same data should be sent over both TTCAN channels), how frequently the data should be updated, *etc.*

Some of the memory blocks are used for communication with the peripheral devices. For these blocks, some additional parameters must be set. In addition to hardware-specific parameters, two valid value intervals can be defined. The first interval defines the expected range of the values produced by an application, and the second one defines the range of the values expected by the hardware device. By these intervals, for example, the values can automatically be scaled into appropriate ranges for A/D or D/A conversion. An application can directly work with abstract physical values independent of the digital ones produced by hardware. A given range also represents the minimum and maximum values of the data; it serves as some sort of limiter and can be used for data validation. As a shortcoming, only linear automatic transformation can be described at this time. More complex transformations can be implemented in C code and added to applications.

To support fault tolerance, it is possible that more than one function block produces the data for some specific memory cell, however, not at the same time. Thus, there can be more than one output memory block with the same ID. To assure that only one replica actually produces the value, the sending block has an additional enable signal, which is used to enable and disable data transmission in different operation modes.

6.4.2 Generation of Time Schedules for the TTCAN Communication Protocol

In time-triggered communication, appropriate time-schedules must be calculated for the messages. In the project, an external tool was implemented for this purpose. Its input information is acquired from the applications as described in the development tool, scanning for the blocks working with variables in the form of shared memory cells. To do this, an additional script was incorporated into the MATLAB®/Simulink®toolset. The script identifies all nodes in a system, all corresponding memory cells, the direction of the information flow, *etc.* The resulting information on all communication paths is written into a specially formatted file, which is read by the external tool for timetable generation. This tool determines the optimum cycle time and the maximum number of lines in the system matrix. Then, the message allocation process tries to determine the optimum message position within the matrix

and the number of columns in the matrix. At the end, the tool tries to opti-
mally re-distribute the remaining free time (*i.e.*, to keep empty message slots
together). Based on the schedules obtained, appropriate tables are generated
(in form of C header files) and added to the generated application code. The
tables serve to initialise the TTCAN communication subsystem. For better
understanding of the results, a simple graphical output in Hypertext Mark-up
Language (HTML) is also generated. An example is shown in Figure 6.10.

Messages

Name	Id	Period	Duration	Origin	Destination
M0_Msg4	536	768	32	M0	S1
S0_Msg1	514	512	32	S0	M0
S1_Msg2	523	672	32	S1	
M0_Msg1	512	1280	16	M0	M1 S0
S0_Msg3	530	1280	16	S0	M1
S1_Msg3	531	1280	16	S1	M0
M0_Msg2	520	1512	32	M0	S1
M1_Msg1	513	1512	32	M1	
M1_Msg2	521	1536	32	M1	M0
M1_Msg3	529	1536	32	M1	M0 S1
S0_Msg2	522	1024	32	S0	M1
M0_Msg3	528	1544	48	M0	M1
M1_Msg4	537	1024	48	M1	S1
S0_Msg4	538	1280	48	S0	M0
S1_Msg1	515	2304	48	S1	M1

System Matrix

Cycle: 512 Transmit: 160 + 304 = 464									
160		32	32	32	16	32	32	32	48	48
	Ref Msg reserved	M0_Msg4	S0_Msg1	S1_Msg2	M0_Msg1	S1_Msg3	M1_Msg1	M1_Msg3	M0_Msg3	S0_Msg4
		M0_Msg4	S0_Msg1	S1_Msg2	S0_Msg3	M0_Msg2	M1_Msg2	S0_Msg2	M1_Msg4	S1_Msg1
		M0_Msg4	S0_Msg1	S1_Msg2	M0_Msg1	S1_Msg3	M1_Msg1	M1_Msg3	M0_Msg3	S0_Msg4
		M0_Msg4	S0_Msg1	S1_Msg2	S0_Msg3	M0_Msg2	M1_Msg2	S0_Msg2	M1_Msg4	

Fig. 6.10. An example of a TTCAN system matrix

The TTCAN standard requires some limitations concerning the matrix
dimensions and the timing of the columns to be considered. Usually several
different schedules can be found for a given message set. The best schedule is
identified by some form of cost function. A possible example of such a function measures the total jitter present in the matrix. Owing to the scheduling
problem's high complexity and all the restrictions imposed by the TTCAN
standard, the common search methods are slow and difficult to use. Evolu-
tionary computing with genetic algorithms proved to be a viable alternative.
In contrast to greedy search or any hill-climbing algorithm, the search through
the solution space is guided by the laws of selection and survival of the fittest.
The search is, therefore, more immune to the local minima problem. With the
genetic algorithms approach the optimisation criteria can be extended to include other aspects of the schedules with minimum time impact. For example,

in order to increase bus bandwidth and to improve fault tolerance, multiple overlapping schedules are highly significant for the IFATIS platform.

6.4.3 Development Process

The process proposed for building fault-tolerant control applications by using the solutions described above is shown in Figure 6.11. First, to develop a control application the model-based approach and the MATLAB®/Simulink® toolset is used without considering the target platform. The resulting model is tested and validated by means of all practices common in the control domain. Later, the model must be prepared for execution in a distributed environment. As mentioned above, to model an application's distributed execution on several processing nodes, the model is divided into submodels. Since all code will be pre-loaded onto the processing nodes prior to execution, all possible execution modes of the system must be considered, *i.e.*, control logic for all different working modes and the decision logic must be put into each subsystem using appropriate Simulink® blocks.

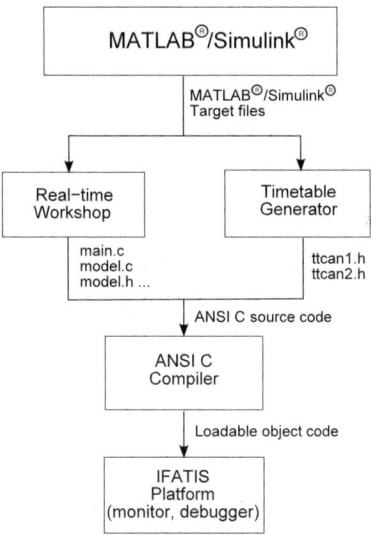

Fig. 6.11. Application development process using the extended MATLAB®/ Simulink® toolset with additional software

To simplify this, each subsystem can be further divided into several smaller ones representing the model of execution in different execution modes. For the decision logic it is most useful if stateflow diagrams are used. Then, shared memory blocks are used to describe the connections between the different subsystems. After that, the application can be simulated and validated again as a distributed system. Sampling effects due to latencies introduced by the

TTCAN bus can also be simulated accurately. Once the results of testing are satisfactory, the specific blocks are augmented with more detailed implementation attributes. Then code is generated and joined with the communication-schedule tables produced in parallel. After compilation, the object code is loaded into the target. Finally, additional tests are performed by using testing and diagnostic tools directly on the target.

7

Asynchronous Real-time Execution with Runtime State Restoration

by Martin Skambraks

This chapter presents the architectural concept of a safety-related programmable electronic system (PES) which combines the benefits, but eliminates the drawbacks of synchronous and asynchronous programming. Thereby, this concept achieves a high degree of fault tolerance through state restoration from redundant PES instances. The PES architecture designed has two key characteristics, namely, *task-oriented real-time execution without asynchronous interrupts* and *state restoration at runtime*.

7.1 Design Objectives

For ease of understanding, the initial objectives that were pursued in the design of the PES concept are to be clarified. They can be subdivided into three categories:

Design for simplicity. Design simplicity prevents engineering errors and, later, eases safety-licensing [78]. That is why one key intention was to strive for minimum complexity by following design for simplicity as a major development guideline. Consequently, minimising design complexity was regarded with higher priority than increasing computational performance. This strategy was pursued not only to lower the cost of safety-licensing, but also to make the application of formal verification methods practically feasible with reasonable effort.

Combining the advantages of synchronous and asynchronous programming. A second objective was to combine the advantages of synchronous and asynchronous programming. Synchronous programming leads to inherently simple PES architecture and temporal behaviour, but its cyclic operating principle limits the field of application. On the other hand, asynchronous programming has a less restricted field of application and is more problem-oriented, but leads to a much more complex PES architecture and temporal behaviour.

Combining the advantages of both concepts arises from the idea to follow the task paradigm of asynchronous programming, but to fragment task execution into discrete cycles of constant length. As for synchronous programming, cyclic execution leads to simple temporal behaviour, but organising application software in independent tasks results in a less restricted field of application. Furthermore, the task paradigm provides a more problem-oriented programming style, since the semantic gaps are prevented that arise, *e.g.*, from mapping time limits of a problem specification to the major cycles of a synchronously operating system.

Unified concept for safety functions. The third objective was to minimise system complexity by integrating a unified concept for safety functions. The fundamental idea is that PESs are configured redundantly, and that each PES instance outputs a Serial Data Stream (SDS) which provides full information about the internal processing. These SDSs were intended to support non-intrusive monitoring and recording of process activities, error detection, and state restoration at runtime. The latter means that, if a PES is in a faulty state due to a transient hardware fault, the SDSs of redundant PESs enable one to copy the internal state and to resume processing at runtime.

Dedicated hardware structures were considered as means to achieve reasonable performance with a minimum of architectural complexity. In other words, a lower architectural complexity was intended to be reached at the cost of a higher number of logic gates. As a result, two fundamental design ideas have been devised: *task-oriented real-time execution without asynchronous interrupts* and *runtime state restoration through serial data streams*.

7.2 Task-oriented Real-time Execution Without Asynchronous Interrupts

The real-time execution concept introduced here combines the advantages of synchronous and asynchronous programming. This is realised by quantising time into discrete *Execution Cycles* of fixed duration. Additionally, the tasks of application software are each partitioned into a number of *Execution Blocks*. These blocks have the following characteristics.

- Each Execution Block is completely executable within a single Execution Cycle.
- The execution of a block is neither interruptible nor pre-emptable.
- The content of the processor registers is lost at the end of any cycle. Thus, data exchange between blocks is only possible *via* the data memory.
- The Execution Blocks of a task are indexed for identification.
- The Execution Blocks of a task do not need to be executed in consecutive order. At the end of an Execution Block, a pointer is generated which identifies the task's block to be executed next.

Task administration and task execution are carried out by two strictly separated units. This allows one to implement the *Task Administration Unit (TAU)*, which is responsible for task state transitions and processor scheduling, in the form of dedicated logic circuitry. The Execution Blocks are processed by the *Application Processing Unit (APU)*, which is based on a conventional microprocessor. Due to its beneficial characteristics in terms of safety and security, a processor with the Harvard architecture is preferable.

7.2.1 Operating Principle

The Execution Cycles consist of two phases, *Execution Phase* and *Administration Phase*. At the begin of the Execution Phase, the TAU outputs the ID of the Execution Block that needs to be processed. Then the APU processes this Execution Block and – after the block has been completed – outputs a pointer to the task's next block. This pointer is read by the task administration functions. If the executed block was a task's last one, *i.e.*, a task's execution has been completed, a distinct pointer value informs about it. The administration functions, which are executed during the Administration Phase, comprise task state control and processor scheduling. At the end of the Administration Phase, the identifier of the Execution Block that needs to be processed in the next cycle is output. Figure 7.1 illustrates this operating principle.

Obviously, the cycle length affects the minimum feasible response time. Hence, a short cycle duration is desirable. On the other hand, the shorter the cycle time the lower is the percentage of time actually spent for task execution, since the time needed for task administration is independent of the cycle duration. Hence, minimising the execution time of the administration algorithms is crucial to realise acceptable real-time performance.

That is why the task administration is preferably implemented in form of a digital logic circuitry. This allows one to speed up execution without algorithmic techniques that lower the computational effort but increase architectural complexity. Additionally, it enables special custom-built opportunities to access the task administration data like, *e.g.*, hard-wired linkage to a digital logic circuit. The latter is essential to integrate the concept of state restoration at runtime introduced later in this chapter.

The task administration follows the scheduling concept of the real-time programming language PEARL [29]; see Section 4.1.2. It provides the most general form of task activation schedules [50]. The hardware implementation of the task administration inherently supports such activation plans. Therefore, each task is assigned a set of parameters, which facilitates configuration for various different activation conditions, *e.g.*, delayed activation on occurrence of an asynchronous event signal.

Most real-time systems follow the approach of rate-monotonic, fixed-priority scheduling. This 'timeless' scheduling policy is merely advantageous for systems that do not provide explicit support for timing constraints, such as

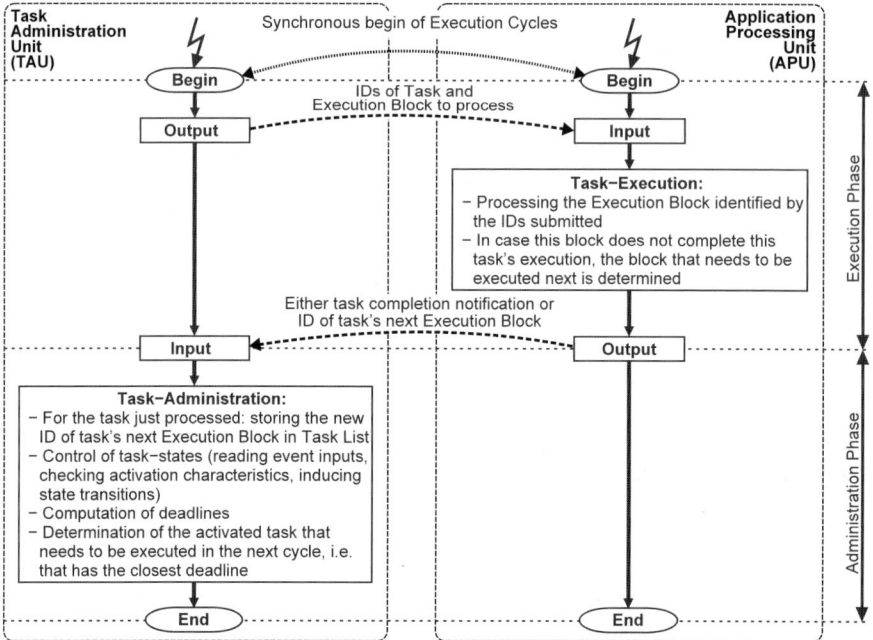

Fig. 7.1. Operating principle: time is quantised into *Execution Cycles* and tasks are processed in discrete *Execution Blocks*

periods and deadlines [12]. Thus, rate-monotonic scheduling is actually inadequate for safety-related systems, for which hard real-time constraints always must be specified exactly. That is why the proposed PES architecture applies the EDF algorithm. Implementing the task administration functions as logic circuit eliminates the differences between rate-monotonic and EDF scheduling in terms of computing time.

The demand for fast execution of the administration algorithms leads to the idea of implementing them in form of a digital logic circuit that executes these kernel algorithms for all tasks in parallel. Unfortunately, even a simplified form of the general task activation plan mentioned above requires, under EDF scheduling, at least two comparisons and two additions of time values per task. Considering the facts that typical real-time applications consist of 10-50 tasks, and that an appropriate binary representation of absolute time values requires at least 24 bits, it is obvious that the hardware expense for such an implementation as a digital circuit is unacceptably high.

For this reason, the hardware architecture of the task administration combines parallel and sequential processing. The kernel algorithms are structured in a way as to allow for parallel processing of the operations related to a single task, whereas all tasks are sequentially subjected to these operations. Figure 7.2 illustrates this processing pattern. It shows the main parts of the

Task Administration Unit (TAU), *viz.*, the *Task List Memory (TLM)* and the *Task Parameter Administration (TPA)*.

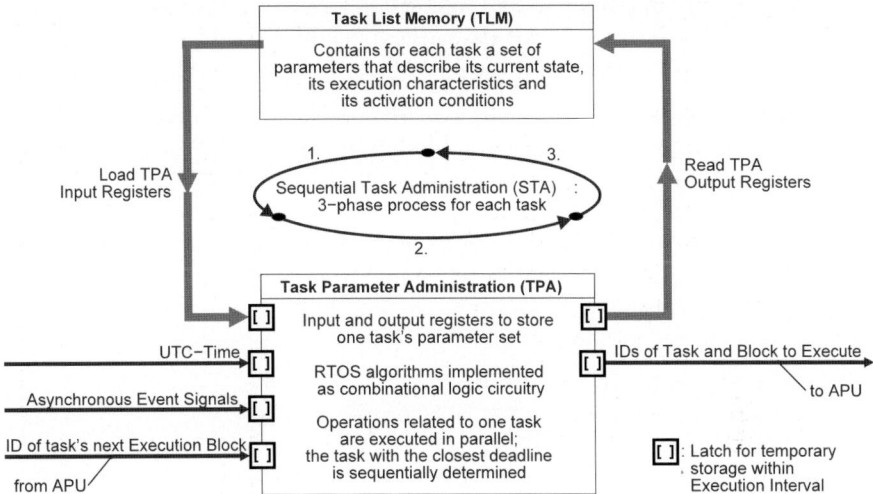

Fig. 7.2. TLM and TPA closely co-operating during *Sequential Task Administration (STA)*

Sequential Task Administration

The TLM administrates a *Task List*, which contains for each task a set of parameters such as its current state, its execution characteristics, and its Activation Plan. The task list has a static size, *i.e.*, all tasks a certain application consists of must be registered at set-up time. Thus, in conformance with the requirements of IEC 61508 for SIL 4 applications, dynamic instantiation of tasks is not supported. Instead, the activation characteristics of a task can be modified.

The TLM co-operates closely with the TPA while sequentially processing all tasks within each Execution Cycle. During this Sequential Task Administration (STA), a three-phase process is initiated for each task:

1. First, the TLM is accessed and the task's entire parameter set is transferred to dedicated input registers of the TPA.

2. Then, the TPA processes the task data and outputs an updated parameter set. This is achieved by purely combinational logic within one clock cycle.

3. During the last phase, the updated task data are transferred back from the TPA to the TLM.

This way, the TPA carries out the following operations in the course of STA:

1. Checking the activation characteristics,
2. Inducing task state transitions,
3. Computing deadlines,
4. Generating updated task parameters,
5. Determining the task with the earliest deadline, and
6. Identifying the Execution Block that needs to be processed next.

The first four operations are separately executable for each task. Therefore, they are performed in parallel by a combinational digital circuit. The fifth item requires to compare the deadlines of all activated tasks. This is carried out sequentially, while the identifier and deadline of the most urgent task is temporarily stored within the iterations of the three-phase process. The ID of the Execution Block that needs to be processed in the subsequent Execution Cycle is output at the next Execution Cycle's begin.

7.2.2 Priority Inheritance Protocol

One of the problems to be solved when designing a real-time task execution mechanism is the priority inversion problem connected with resource sharing (*cf.* also Section 2.2.1). The problem arises when a lower priority task seizes a resource and is pre-empted by a higher priority task who also requests the same resource. The latter task thus attains the same (low) priority of the former one which is blocking the resource.

The problem can be prevented by the *Priority Inheritance Protocol (PIP)* method. Realising mutually exclusive access to shared resources requires to associate each resource with a binary semaphore (see Section 2.3.2), and to provide the two basic semaphore operations 'Request' and 'Release'.

The PIP can be directly implemented together with the two semaphore operations. For this, the Request operation must include mechanisms for priority inheritance and the Release operation for restoring the original priority. Both operations must be non-interruptible, since inconsistencies like, *e.g.*, a semaphore is released but the original priority has not yet been restored are unacceptable. Furthermore, the scheduling algorithm has to be executed afterwards to take the updated priorities, *i.e.*, deadlines in case of EDF scheduling, into account. This demand for non-interrupted execution followed by running the scheduling algorithm can easily be combined with the proposed cyclic operating principle. In fact, integrating the PIP is straightforward.

The PIP requires semaphores to be requested in a nested fashion to avoid deadlocks [106]. For this, the semaphores are ordered by identification numbers. A semaphore m' is said to be of higher order than semaphore m if it has the higher identification number, *i.e.*, $m' > m$. If a task requires multiple resources at the same time, the associated semaphores must be requested in decreasing order and released *vice versa*.

Semaphore Parameters

In order to implement the PIP into the proposed PES architecture, the following parameters must be administrated for each semaphore m:

$S_{State}(m)$:   This variable stores the state of semaphore m. The state is either *free* or *occupied*.
$$S_{State}(m) = \{free\,|\,occupied\}$$

$S_{Task}(m)$:   While semaphore m is in the state *occupied*, this parameter stores the number of the task holding semaphore m.

$S_{Deadline}(m)$:   While task i holds semaphore m, this parameter stores the deadline that is restored when the task i releases the semaphore. The parameter is represented by an absolute time value. At the moment a semaphore m is assigned to a task, $S_{Deadline}(m)$ is assigned the current deadline of that task. Subsequently, $S_{Deadline}(m)$ is only modified if semaphore m inherits a deadline d and the associated task $S_{Task}(m)$ also holds an higher-order semaphore m' that has the inherited deadline d, too. This ensures that the deadline d is restored when semaphore m is released, which is required since semaphore m' is still held by $S_{Deadline}(m)$.

$S_{NxtLwr}(m)$:   While task i holds semaphore m, this parameter indicates the next lower-order semaphore that task i requested, independent of the fact whether task i has assigned that semaphore or is blocked by it. In case the task holding semaphore m does not hold any lower-order semaphore, a special value informs about it.
$$S_{NxtLwr}(m) = \{no\,lower\,order\,sema\,requested\,|\,\#j\}, \quad j \leq m$$

$S_{Blocked}(m)$ While semaphore m is in the state *occupied*, this set contains the numbers of all tasks blocked by it.
$$S_{Blocked}(m) = \{i : (\text{task } i \text{ is blocked by semaphore } m) \wedge i \neq S_{Task}(m)\}$$

The formal description of the Request and Release operations requires one further parameter:

$L(i)$ is a set that contains the numbers of all semaphores that task i holds. As long as a task holds no semaphore, the parameter is an empty set $\{\}$. Any time a task is assigned a semaphore m, the number of that semaphore is added to the set $(L(i) := L(i) \cup \{m\})$. When a task releases semaphore m, its number is removed from the set $(L(i) := L(i) \setminus \{m\})$. The number of the lowest-order semaphore held by a task is $\min\{L(i)\}$.

The information about the semaphores that a task has already requested when performing a Request or Release operation is implicitly given by the program code, *i.e.*, $\min\{L(i)\}$ does not require one to search for a minimum number

```
Request(m) :

        if  (L(i) ≠ {}) then
            S_NxtLwr(min{L(i)}) := m;
        end if;
        if  (S_State(m) = free)   then
            S_State(m) := occupied;
            S_Task(m) := i;
            S_Deadline(m) := P_Deadline(i);
            S_NxtLwr(m) := no lower order sema requested;
            L(i) := L(i) ∪ m;
        else
            S_Blocked(m) := S_Blocked(m) ∪ {i}
            P_Deadline(S_Task(m)) := min {P_Deadline(S_Task(m)), P_Deadline(i)};
            m' := m;
            m'' := S_NxtLwr(m);
            while (m'' ≠ no lower order sema requested) do
                P_Deadline(S_Task(m'')) := min {P_Deadline(S_Task(m'')), P_Deadline(i)};
                if (S_Task(m') = S_Task(m'')) then
                    S_Deadline(m'') := min {S_Deadline(m''), P_Deadline(i)};
                end if;
                m' := m'';
                m'' := S_NxtLwr(m'');
            end while;
            suspend(i);
        end if;
        return;
```

Fig. 7.3. Pseudo-code of the Request operation

in a set – the number of the lowest-order semaphore held by a task is known at compilation time. Hence, the set $L(i)$ does not need to be handled by the program code it is only necessary for the subsequent pseudo-code descriptions of the two commands.

Request Operation

Figure 7.3 shows the pseudo-code of the Request algorithm. Its first action depends on whether the requesting task i already holds a semaphore, *i.e.*, the set $L(i)$ is not empty. If this is the case, the parameter S_{NxtLwr} of the lowest semaphore held by task i is assigned m, *i.e.*, $S_{NxtLwr}(\min\{L(i)\}) := m$. Then, the algorithm checks whether semaphore m is already held by another task.

If the semaphore is available, *i.e.*, $S_{State}(m) = free$, its state is transferred to *occupied* and the parameter $S_{Task}(m)$, which indicates the holding task, is set to i. Additionally, the parameter $S_{Deadline}(m)$, which stores the deadline that is restored after release, is set to the current deadline of task i, and the parameter $S_{NxtLwr}(m)$, which points to the next lower-order semaphore held by task i, is assigned to indicate that task i has no lower-order semaphore requested. The line $L(i):=L(i)\cup m$ indicates that m now belongs to the set of semaphores held by task i.

If the semaphore is already allocated, i.e., $S_{State}(m)\neq$ free, the number of the requesting task is added to the set of tasks blocked by semaphore m, i.e., $S_{Blocked}(m):=S_{Blocked}(m)\cup\{i\}$. Then, the algorithm checks whether the deadline of the requesting task is closer than the deadline of the task holding m, i.e., $P_{Deadline}(i)<P_{Deadline}(S_{Task}(m))$. If this is the case, the task holding m inherits the deadline from the task requesting that semaphore, i.e., $P_{Deadline}(S_{Task}(m)):=\min\{P_{Deadline}(S_{Task}(m)),P_{Deadline}(i)\}$.

After that, the algorithm checks whether transitive priority inheritance is necessary. For this, a while loop is applied to follow the chain of semaphores defined by the pointers $S_{NxtLwr}(m)$ until a semaphore $m^{(x)}$ is reached, whose pointer $S_{NxtLwr}(m^{(x)})$ indicates that the holding task $S_{Task}(m^{(x)})$ has no lower-order semaphore requested. For each of the semaphores that belong to that chain, the algorithm checks whether the deadline of the task holding it is closer than the deadline of the requesting task i. If so, the task holding that particular semaphore inherits the deadline from task i.

Further action is necessary if a task holds more than one semaphore of the chain. This is the case if semaphore m'' inherits the deadline $P_{Deadline}(i)$ and the associated task $S_{Task}(m'')$ also holds a higher-order semaphore m' that has already inherited deadline $P_{Deadline}(i)$. In this case, the condition $S_{Task}(m')=S_{Task}(m'')$ is valid. Hence, if the deadline $P_{Deadline}(i)$ is closer than $S_{Deadline}(m'')$ of the lower-order semaphore m'', then $S_{Deadline}(m'')$ is also assigned $P_{Deadline}(i)$. This ensures that the deadline $P_{Deadline}(i)$ is restored when semaphore m'' is released, which is required, since – at that moment – semaphore m' is still held by the task that just released m''.

The command suspend(i) is the last to be executed if semaphore m has already been allocated at the time of request. This command, which submits a suspend operation to the task administration, has to be the last command of an Execution Block. The rest of the Request algorithm is part of the next Execution Block. Hence, task i is suspended before the Request algorithm completes. Task i will return to the calling program part after semaphore m has been released and assigned to task i.

Release Operation

The pseudo-code of the Release operation is depicted in Figure 7.4.

When task i releases semaphore m, the semaphore does not belong anymore to the set of semaphores held by the task i, i.e., $L(i):=L(i)\setminus\{m\}$. If task i holds another semaphore with a higher order than m, i.e., $L(i)\neq\{\}$, the pointer S_{NxtLwr} of the lowest-order semaphore $\min\{L(i)\}$ is set to a value indicating that task i has no lower order semaphore requested.

Then the algorithm replaces the current deadline of the releasing task i by the value of $S_{Deadline}(m)$, i.e., $P_{Deadline}(i):=S_{Deadline}(m)$. If task i does not hold a higher-order semaphore that inherited a deadline d, this ensures that

```
Release(m) :
```
$L(i) := L(i) \setminus \{m\}$;
if $(L(i) \neq \{\})$ **then**
   $S_{NxtLwr}(\min\{L(i)\}) := no\ lower\ order\ sema\ requested$;
end if;
$P_{Deadline}(i) := S_{Deadline}(m)$;
if $(S_{Blocked}(m) = \{\})$ **then**
   $S_{State}(m) := free$;
else
   $i' : P_{Deadline}(i') = \min\{P_{Deadline}(k) : k \in S_{Blocked}(m)\}$;
   $S_{Blocked}(m) := S_{Blocked}(m) \setminus \{i'\}$
   if $(L(i') \neq \{\})$ **then**
      $S_{NxtLwr}(\min\{L(i')\}) := m$;
   end if;
   $S_{State}(m) := occupied$;
   $S_{Task}(m) := i'$;
   $S_{Deadline}(m) := P_{Deadline}(i')$;
   $S_{NxtLwr}(m) := no\ lower\ order\ sema\ requested$;
   $L(i') := L(i') \cup m$;
   `continue(i)`;
end if;
`return`;

Fig. 7.4. Pseudo-code of the Release operation

the original deadline is restored. Otherwise, the assignment restores the deadline d, which has been copied to $S_{Deadline}(m)$ due to transitive inheritance.

If the set of tasks blocked by m is empty, i.e., $S_{Blocked}(m) = \{\}$, the semaphore state is transferred to *free*. Otherwise, the release algorithm determines the task i' of $S_{Blocked}(m)$ with the closest deadline and removes that task from the set of tasks blocked by m, i.e., $S_{Blocked}(m) := S_{Blocked}(m) \setminus \{i'\}$.

If task i' already holds a semaphore, i.e., the set $L(i')$ is not empty, the parameter S_{NxtLwr} of the lowest semaphore held by task i' is assigned m, i.e., $S_{NxtLwr}(\min\{L(i')\}) := m$.

The subsequent five assignments equal the branch for $S_{State}(m) = free$ of the Request algorithm. Since semaphore m is still in the state *occupied*, the transfer to this state is not really necessary. Nevertheless, the pseudo-code contains the state assignment $S_{State}(m) := occupied$ for clarity. The parameter $S_{Task}(m)$ is set to i', the parameter $S_{Deadline}(m)$ is set to the current deadline of task i', and the parameter $S_{NxtLwr}(m)$ is assigned a value indicating that task i has no lower-order semaphore requested. The line $L(i') := L(i') \cup m$ indicates that m now belongs to the set of semaphores held by task i'.

Finally, a `continue(i')` operation is submitted to the task administration before the system returns to the calling program part of task i.

Integration into the Cyclic Operating Style

Both algorithms must be executed without interruptions and, hence, they must be implementable into one Execution Block (except the commands that follow the `suspend(i)` operation of the Request algorithm, which must be implemented in a succeeding Execution Block). This defines a minimum length for the Execution Blocks.

As already mentioned, the syntax $\min\{L(i)\}$ is only used to formally describe these algorithms, *i.e.*, the lowest-order semaphore held by task i is known at compilation time and does not need to be searched. A significant number of computation steps is caused by the `while`-loop of the Request algorithm, and – in terms of the Release algorithm – by the search for the blocked task with the closest deadline. Typically, resources are shared by a few tasks and transitive priority inheritance is possible *via* a limited number of tasks, only. This limits the computational load caused by the `while`-loop and the search for the closest deadline. Consequently, this characteristic is valuable to define the minimum length of the Execution Blocks.

7.2.3 Aspects of Safety Licensing

Following the task concept of asynchronous programming but operating in discrete cycles similar to the synchronous approach provides a very problem-oriented programming style, since semantic gaps, *e.g.*, from mapping time limits of a problem specification to the major cycles of a synchronously operating system, are prevented from arising. Unlike synchronous programming, the proposed concept does not restrict the realisation of process-dependent program flows and, thus, is applicable to a large field of applications. Some significant characteristics in terms of safety licensing are *compliance with IEC 61508*, *simple temporal behaviour* and *low effort for verification*.

Compliance with IEC 61508

The operating principle renders the use of asynchronous interrupts superfluous. Additionally, the hardware-implemented task administration physically restricts the task activation to one instance at a time, and does not support dynamic instantiation of tasks. In comparison to conventional task-oriented real-time systems, this provides a higher conformity with the safety standard IEC 61508, which restricts the use of interrupts and prohibits dynamic instantiation of objects for applications of highest safety criticality.

Handling of interrupts and multiple activation instances of tasks as well as dynamic task instantiation involves dynamic memory usage, *e.g.*, stack memory. Although the associated algorithms are, in general, not difficult to implement, their formal verification tends to be very intricate. In particular, formal proofs that no memory overrun can occur are complex and, hence, the confidence required for SIL 3 and SIL 4 applications is difficult to achieve –

if not impossible. Actually, this is why IEC 61508 highly recommends that these methods are not used. Thus, the higher compliance with IEC 61508 is closely related to lower efforts for formal verification.

Simple Temporal Behaviour

The task processing in discrete Execution Cycles makes further distinction between interruptible and non-interruptible program sections superfluous. This not only has a positive effect on feasibility analysis; it also lowers the temporal behaviour's complexity in general. The minimum feasible response time can be computed by simply adding the maximum release jitter and the cycle duration. Moreover, the WCET of a task can very easily be obtained by counting the Execution Blocks that must be processed following the longest path of a task instance. Furthermore, during fragmentation of program code into Execution Blocks, the performance gain due to cache-memory structures can the taken into consideration by assuming an empty cache at any cycle's beginning.

For conventional task-oriented real-time systems, determining the minimum feasible response time or a task's WCET is far more complex, since all potential interruptions must be taken into consideration. Not only is the maximum delay due to context switches difficult to analyse, but also their influence on performance due to changes of cache-memory content. Of course, the worst-case scenario is – theoretically – determinable, since the maximum number of interruptions is given by the task execution characteristics, *e.g.*, the tasks' activation frequencies. However, the algorithms that search and investigate the worst-case scenario have significant complexity, in particular, since these algorithms must take the concurrent behaviour of all tasks as well as an asynchronously running real-time operating system into account. As a result, formal verification of these algorithms takes enormous effort – if it is possible at all.

The complexity of these algorithms can be reduced at the price of performance. For instance, performance gains due to cache memories become determinable by setting up non-interruptible code sections, but this approach has a negative impact on feasibility analysis.

In contrast to the algorithms that investigate worst-case scenarios in conventional systems, the algorithms splitting the program code of a task into Execution Blocks operate on a single task, *i.e.*, they do not need to take concurrency of all tasks into account. Hence, these algorithms have lower complexity and are easier to verify, although the fragmentation of program code also requires to investigate worst-case execution times.

Low Effort for Formal Verification

Implementing the task administration functions as digital logic circuitry allows one to accelerate their execution through parallelism and pipelining. As a result, they can be completely executed within any Execution Cycle without causing unreasonably high computational load or unacceptable response times. In other words, there is no need to decrease the minimum feasible response time by splitting the kernel functions into multiple layers, or by handling lists of activated and suspended task as conventional real-time operating systems frequently do, *cf.* those introduced in [50, 75]. That is why the proposed concept leads to a particularly low architectural complexity, which, in turn, eases formal verification.

7.2.4 Fragmentation of Program Code

The considered operating principle requires one to divide the program code of a task into fragments that are completely executable within an Execution Cycle. These code fragments are called Execution Blocks. Each one has a unique identification number assigned, and at the end of a block a pointer to the next Execution Block is generated. As Figure 7.5 illustrates, this concept allows arbitrary program flows.

On its left side, the figure presents a typical task algorithm in the form of a program flowchart. The subdivision into Execution Blocks is illustrated in the middle. Sequential program parts are simply split into a number of code fragments that are executable within an Execution Cycle. An example for this is command block 2, which is divided into five Execution Blocks. Loops and conditional branches are realised by pointer alternatives at the end of an Execution Block. As visible in the figure, command block 1 is implemented together with conditional branch 1 in four Execution Blocks, and the last block realises the branch by pointing to two alternative successor blocks. In case a loop or a conditional branch can be completely executed within one Execution Cycle, it is implemented in one Execution Block. As an example, command block 4 and conditional branch 3 are integrated into one single Execution block. The right part of Figure 7.5 presents two possible paths through the program code.

This operating style leads to three questions:

- How does the concept restrict software development?
- How can a task's program code automatically be divided into Execution Blocks?
- How does fragmentation affect computing performance?

Restrictions to Software Development

As the example of Figure 7.5 shows, fragmented task execution in discrete cycles facilitates conditional branches and loops. In analogy to the command cycle of a microprocessor, an Execution Block can be considered to be a huge macrocommand which consists of a number of small commands. Hence, the operating style provides Turing-completeness just like any other microprocessor (here, indefinitely enlargeable storage is assumed, as it is usually done when a

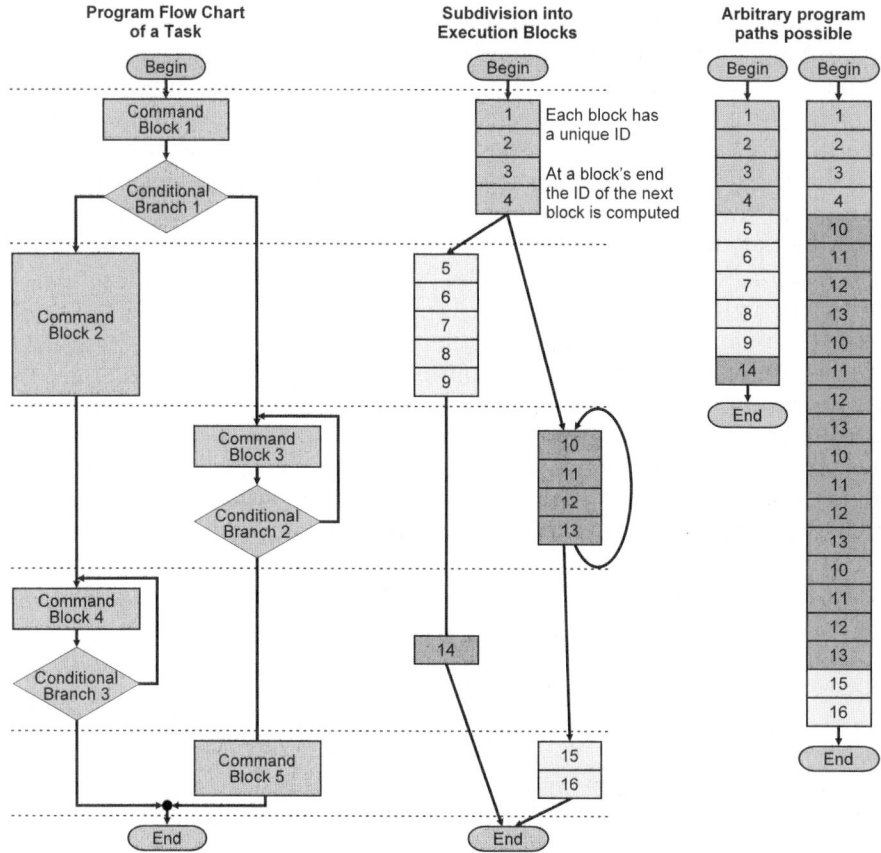

Fig. 7.5. The operating principle facilitates arbitrary program flows

physical system is declared to be Turing-complete). Consequently, the system can perform any calculation that any other computer is capable of, provided enough memory is available. In other words, the cyclic operating style does not restrict the software development in general, it just affects performance. All conventional programming paradigms like, *e.g.*, While-loops, For-loops, If-Then-Else-structures, Case-structures and procedure calls are applicable. This is also true for higher-order paradigms that can be constructed with the help of the former ones, like, *e.g.*, exception handling.

The realisation of conditional branches, which are the underlying technique for While-loops, For-loops, If-Then-Else-structures and Case-Structures, has already been illustrated in the example above. That is why only the implementation of procedure calls needs further explanation. The dynamic, unforeseeable processor assignment of the execution concept does not facilitate the use of a single stack memory accessed by processor functions like Push and Pop. Instead, a separate stack for each task would be necessary to ensure data consistency. Of course, this form of independent stack handling could be realised by software or dedicated hardware. However, since pointers (*i.e.*, stack pointers) and dynamic variables (*i.e.*, stack memories) are common sources for design faults and – what is even worse – most difficult to formally verify, the safety standard IEC 61508 forbids their use in applications that belong to the safety classes SIL 3 and SIL 4. That is why stack memory handling has been considered inappropriate for the proposed safety-related real-time PES.

Procedure calls can be realised without stack memory simply by storing the *jump-back* address, at which a program must be continued after completion of a procedure, in predefined memory cells. The 'return' command is replaced by an unconditional jump to this temporarily stored address. Obviously, a programming paradigm only poorly supported by this technique is recursive procedure invocation, since it requires to reserve for each task as many memory cells for the storage of jump-back addresses as procedure instances can occur simultaneously. However, like pointers and dynamic variables, recursion is a common source for failures, difficult to (formally) verify and – as a consequence – IEC 61508 forbids its use in SIL 3 and SIL 4 applications. Hence, poor support of recursion can be considered an insignificant restriction, and it is sufficient to arrange for the storage of just one jump-back address for each task that calls a procedure.

Automatic Division into Execution Blocks

The fragmentation into Execution Blocks can automatically be performed by compiler software and, hence, the cyclic execution concept imposes no special constraints on source code. In other words, the software developer can write the program code of any task as for conventional real-time programming languages like Ada or PEARL. Of course, the EDF scheduling policy affects programming, since timing constraints must be specified instead of task priorities.

The automatic fragmentation method introduced below founds on the machine code produced by a compiler. Furthermore, the fragmentation bases on the approach that the context of all processor registers is saved at an Execution Block's end and restored at a block's begin. Of course, taking code fragmentation into account already during compilation would provide more opportunities for code optimisation, especially if the temporal storage of the processor register content would be restricted to the values that are used later on. This topic has been considered interesting for future research, but here we assume compilation and fragmentation to be separate. The automatic fragmentation operates as follows.

During an initial program code analysis, all jump destinations are identified and labeled. Then, the program code is sequentially divided into Execution blocks, starting at the top of the program. Thereby, all commands are processed step by step. As long as the program code is a straight order of commands, it is just split into appropriate code fragments. Special action is necessary when one of the previously labeled jump destinations, a program branch, or a procedure's end is reached:

- A jump destination can be the jump-back address of a For- or a While-loop, the address of a procedure's first command, the address for program continuation after a procedure's completion, or one of the jump alternatives of a conditional branch. The latter can be the rejoining point of multiple program paths, *i.e.*, the address of the first command that must be executed after two (or more) program paths rejoined. For each jump destination, a new Execution Block is created whose first command is the one to which the jump destination points. Subsequently, the automatic fragmentation links these blocks' identifiers with their respective jump destination labels.
- A program branch is a conditional jump. It can be the foot point of a For- or a While-loop. Both path alternatives of a conditional jump are handled as jump destinations. The automatic fragmentation processes both branches successively, up to the point where they merge again. This point is denoted by a jump destination that both branches eventually reach. A program branch terminates an Execution Block and the pointer to the next Execution Block is used for the branching. In dependence on the actual program flow, this pointer is set to one of the jump destination labels.
- A procedure's end terminates an Execution Block. The pointer to the next Execution Block is set to the jump-back identifier which has been stored upon calling. The jump-back identifier points to the Execution Block that must be executed when the procedure has been completed. Thus, the first command of this block is the one that sequences the procedure call.

After the entire program code has been processed, the automatic fragmentation replaces the labels of the jump destinations by the corresponding Execution Block identifiers.

The algorithm described so far potentially causes an unnecessary high percentage of idle processing. For example, a program branch is transferred to at least three Execution Blocks, even if the conditional jump and both branches could be executed within one block. Hence, only a small fraction of the blocks is used for actual program code processing and the rest of the blocks' available execution time is filled with idle processing. That is why the automatic fragmentation ends with an optimising phase, which merges program fragments that fit into one Execution Block.

The optimisation founds on an analysis of all branches. For any branch, the point is determined at which its alternative program paths merge again. In case even the longest path of a branch can be processed within one Execution Cycle, the branch and its paths are integrated in a single Execution Block. The optimisation also combines all Execution Blocks of a loop into one block, provided even the maximum iteration number can be completed within one Execution Block. Once a loop or a branch has been merged to a single Execution Block, the optimiser tries to merge it with preceding or consecutive blocks. This optimisation procedure is carried out iteratively until no further blocks can be merged.

Figure 7.6 illustrates the optimisation. The program flow example of the figure contains a loop with two alternative branches. The initial fragmentation, which creates one Execution Block for each jump destination, returns four blocks. In its first step, the optimisation discovers that the two alternative branches fit into one block. After the associated Execution Blocks (2 and 3) have been merged to one Block (23), the optimisation checks whether this new block can be merged with the preceding and the succeeding blocks. Since this is possible, the initial four blocks can be merged to one block.

Depending on the size of a procedure, it might be beneficial to replace a procedure call, which always instantiates at least two Execution Blocks, by its source code. As Figure 7.7 shows, this allows one to integrate a procedure in the calling block. Especially for short procedures this can result in a considerable reduction of Execution Blocks. However, the gain in computing performance – which results from the lower number of blocks that must be processed for task completion – is paid for by an increased demand for program memory, since multiple instances of a procedure must be stored. That is why the optimising algorithm merges procedures with their calling blocks only if it results in a shorter WCET.

Figure 7.7 illustrates the merging. The figure contains a program flow example with two calls of the same procedure. The initial fragmentation contains only one instance of this procedure, which consumes one Execution Block (5). During the subsequent optimising steps, the procedure calls are replaced by the procedure's source code. Thus, the source code is implemented twice, depicted by the notation 5a and 5b. As a result, the number of Execution Blocks that needs to be processed on the longest program path, *i.e.*, the WCET, is reduced.

Influence on Computing Performance

There is no doubt that the real-time execution concept described strongly limits the achievable computing performance. However, in terms of safety-related systems, computing performance is a minor issue in comparison with safety.

Let C be the computing power of the embedded processor itself. Then, the influence of the execution principle can be modeled by the equation

$$C' = C \times \frac{t_{Exec} - t_{Store} - t_{Restore}}{t_{Cycle}} \times (1 - p_{idle}) \;, \qquad (7.1)$$

where $t_{Restore}$ and t_{Store} are the amounts of processor time needed to restore and store the processor register content at the begin and the end of an Execution Block, respectively. The average percentage to which the Execution Blocks cause idle processing is denoted by p_{idle}. This average is computed for the longest program path, as this reflects the worst-case scenario with the

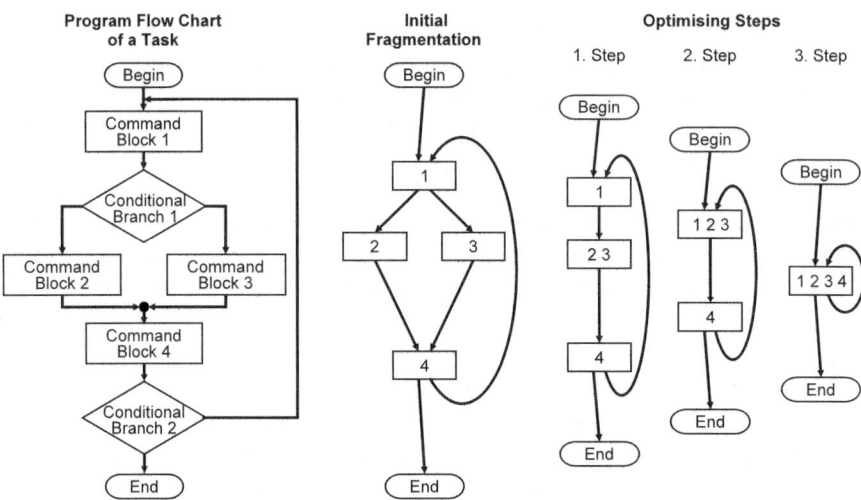

Fig. 7.6. To increase performance, automatic fragmentation tries to combine Execution Blocks during a final optimisation stage

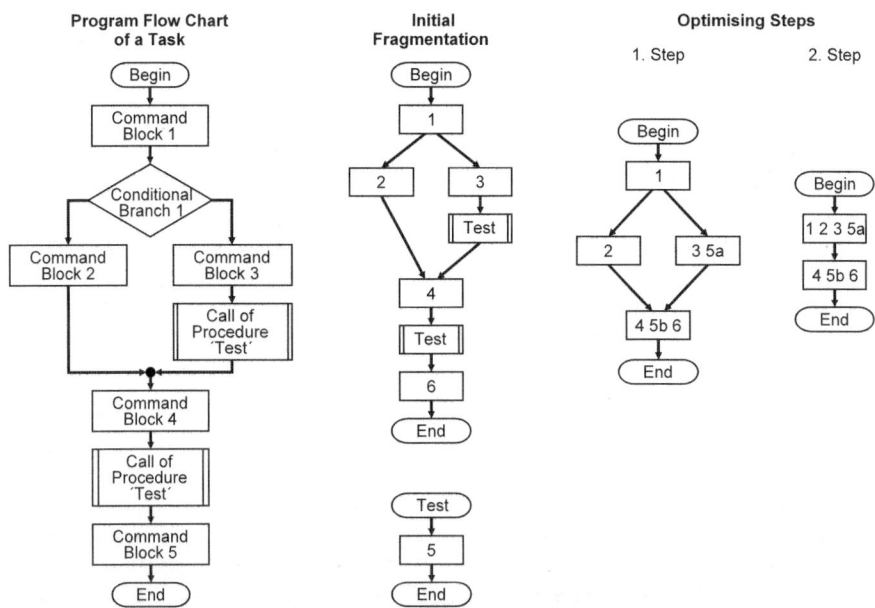

Fig. 7.7. During optimisation, procedure calls might be replaced by their source code. This can reduce the number of Execution Blocks that needs to be processed on the longest program path

longest execution time. Obviously, p_{idle} strongly depends on the cycle duration t_{Cycle}. The longer t_{Cycle}, the more often the Execution Blocks must be filled up with idle processing. On the other hand, a shorter cycle duration causes higher overhead for storage and restoration of the register content. The type of processor and compiler is also assumed to have a considerable influence on p_{idle}, since a more extensive instruction set can provide higher computational power within the same amount of clock cycles. In any case, p_{idle} is less than 50%, because the added processing times of any two successive Execution Blocks must exceed at least the processing time granted to one block. Otherwise, these two blocks could be merged in one block.

Asynchronous Programming:

A direct comparison of the performance achievable with conventional task-oriented real-time systems is problematic, as these can provide higher performance on average, but feasibility analysis restricts their usage. Feasibility analysis requires one to take all possible context switches into account, which not only require one to store and restore the processor register content, but also affect the performance gain due to architectural features such as cache memories or superscalar processing and, hence, make it difficult to compute WCETs. Sometimes the performance gain due to such architectural features is even totally annihilated during feasibility analysis. Moreover, execution interdependencies requiring mutual exclusion are often difficult to discover and frequently unknown. As a consequence, the worst-case scenarios assumed are frequently over-pessimistic, *i.e.*, they potentially contain more context switches than actually possible.

As a result, there is a huge discrepancy between assumed and real worst-case scenarios, but only the latter is relevant for hard real-time systems. The situation improves only slightly by dividing program code into interruptible and non-interruptible segments, since this causes release jitter which, in turn, has a negative impact on feasibility.

Assume a system of the considered architecture with the cycle period t_{Cycle}. A conventional system with a comparable minimum achievable response time would require a fine granularity of non-interruptible program segments with the maximum segment length t_{Cycle}. Adding the fact that this system would have to execute the real-time operating system kernel algorithms on the processor instead of taking advantage of a dedicated hardware, we assume that – in terms of worst-case feasibility – the attainable computing performance is for both systems nearly identically affected by task administration and context switches. Thus, together with the rather pessimistic assumption that p_{idle} is 50% on the longest program paths (*i.e.*, in the worst-case scenario), it can be stated that the cyclic operating principle provides at least 50% of the performance of a conventional system. Using highly sophisticated algorithms for code fragmentation, we assume that the performance can be increased to a level above 70%.

Synchronous Programming:

   In the case of synchronous programming, feasibility is implicitly analysed when an execution schedule is defined. The WCETs of the scheduled blocks are easy to compute. Since their execution cannot be interrupted, there are no unforeseeable context switches, and the performance gain due to architectural features like cache memories or superscalar processing is easy to predict. The schedule of a frame must guarantee, however, that all its scheduled blocks complete within the minor cycle time T_{minor}. Thus, the WCETs of a frame's scheduled blocks must be taken into account as a whole – even if it is unlikely that all processes consume their WCETs simultaneously, *i.e.*, within the same cycle. Together with the fact that aperiodic actions must be realised by cyclic polling, this potentially causes considerable idle processing. The WCET of an aperiodic action must be taken into account for every polling cycle, even if it occurs only rarely (This is why aperiodic actions are often handled by periodic servers, an approach that eliminates the advantage of simple temporal behaviour). The longer the minor cycle period is, the smaller number of frames can fall between two polling instances for a given maximum response time. As a consequence, the performance loss associated with polling is larger the longer the minor cycle period.

   Assume a system of the considered architecture with the cycle period t_{Cycle}. A synchronous system with a comparable minimum achievable response time would require a minor cycle length of t_{Cycle}. Thus, the WCET of any block scheduled must be shorter than t_{Cycle}, and the granularity of code fragmentation is similar for both operating policies.

   In summary, comparing the computing performance of synchronous programming and the considered approach is difficult – and it may depend on the application which operating principle is more advantageous. It cannot be claimed that synchronous programming clearly provides higher performance. Most likely, systems with identical cycle periods will provide nearly the same performance. If the minor cycle period of a synchronous system is larger than the execution cycles of the considered approach, a performance gain is achieved due to less fragmentation and better usage of architectural acceleration features. This gain is compensated, however, by the drawback of cyclic polling, which potentially causes more idle processing for larger periods T_{minor}.

7.3 State Restoration at Runtime

This section introduces the concept of state restoration at runtime, which is applied to tolerate spontaneous hardware failures. This safety function, which can be considered as a form of forward recovery (see also Section 4.1.7), enables a redundant configuration of PESs to bring a failed PES back on line by equalising its internal state with that of its running counterparts. Thus, redundancy attrition due to transient processing faults is prevented. The technique also increases robustness against permanent failures, since it enables one to replace error-affected PESs without the need to interrupt on-going processing. The concept of cyclically operating real-time execution described in the previous section aims to prevent design faults by simplifying both architecture and operating principle. In order to guarantee safe and reliable operation, these strategies must be integrated into a holistic concept, which also takes spontaneous hardware failures into account and does not jeopardise simplicity.

Part 3 of the safety standard IEC 61508 recommends various techniques to reduce the influence of spontaneous hardware failures. Some of these techniques are, *e.g.*, the use of multiple processing units combined with majority voting (IEC 61508-3, Table A.2), double-RAM with hardware or software comparison and read/write test (IEC 61508-3, Table A.6), and RAM-monitoring with a modified Hamming code or other Error Detection/Correction Codes (IEC 61508-3, Table A.6). Unfortunately, none of the techniques recommended covers all possible sources of failures and, as a consequence, several techniques usually need to be combined to attain a required degree of fault tolerance. This potentially increases system complexity and, therefore, somehow contradicts the major design guideline simplicity.

This is the reason why a rather exceptional but more holistic approach has been devised. Instead of processing redundant information inside each PES, the PES itself is designed to be redundantly configured. Each PES instance outputs a *Serial Data Stream (SDS)*, which continuously informs about its internal processing. These data streams, which are organised in *Transfer Cycles* that match the Execution Cycles, allow one to determine the internal states of the PESs exactly. Within a redundant configuration of PESs, the SDSs are exchanged to serve three safety functions:

Non-intrusive Monitoring and Recording of Process Activities. The SDS technique enables to monitor and record process activities with external devices. Since these recording devices are not part of the PES itself, they do not necessarily need to fulfill the same high safety requirements.

Detection of Processing Errors. Each PES can detect processing errors by comparing its SDS with the SDSs of its redundant counterparts.

Forward Recovery at Runtime. If a PES is affected by a transient hardware fault or even completely replaced, the SDSs of the redundant PESs enable to copy the internal state and to re-join the redundant processing at runtime.

All PESs synchronise their operation to the same global time reference UTC. This ensures that the SDSs of redundant PES instances are identical, provided no instance is impaired due to processing errors. The fact that the SDSs of properly operating, redundant PESs are identical enables one to determine the correct SDS values simply by majority voting. For this, each PES includes a *Majority Voter*, which compares the SDSs of all redundant PES instances – including the PES's own SDS. The majority voter uses only *valid* streams to determine the majority. An SDS is denoted as 'invalid' if, *e.g.*, the associated PES is in 'Error State' or if the reception fails. The majority value generated by the voter is used for state restoration and to detect processing errors, *i.e.*, a PES's state is considered to be erroneous if its own SDS differs from the majority.

The SDS concept renders the employment of further fault tolerance measures inside a PES superfluous. Certainly, applying redundancy inside each PES could further increase the degree of fault tolerance, but, since redundancy attrition due to transient processing faults is prevented, this gain is also easily attainable by increasing the number of redundant PES instances. The latter approach has the advantage that it does not increase architectural complexity.

7.3.1 State Restoration at Runtime and Associated Problems

State restoration at runtime requires a redundant configuration of uniformly operating PESs. Each PES must be able to:

- Detect processing errors by comparing its own results with the results of its redundant counterparts,
- Drop out of redundant operation in case of processing errors,
- Copy the internal state of the redundant units at runtime, and
- Re-join redundant operation when state equivalence has been reached.

In the literature, such restoration techniques are frequently called 'forward recovery' or 'roll-forward recovery'. These terms are something of a misnomer, since the literal opposite 'backward recovery' refers to the re-start at special recovery points which have been stored in process history. In analogy, forward recovery could be misinterpreted as a re-start at distinct future points. This would require an estimation of future process behaviour, which is certainly impossible. That is why the phrase 'state restoration at runtime' is used here.

In general, the state of a PES is defined by the content of all its storage elements, *e.g.*, registers, latches, and memory blocks. Consequently, a change of a storage element's content equals a change of the PES's state. Nevertheless, the term *State Change* is avoided in this chapter as it is open to misinterpretation. The term suggests that there exists an explicit model of a finite state machine, which is not the case. Such models are usually created at the application level, since a model on the hardware level of a PES would have an inappropriate extension. That is why the term *Data Modification* is used instead.

In the sequel, the state of a PES is regarded to be only defined by those storage elements that can – either directly or indirectly – influence future processing. These storage elements, which can comprise multiple memory blocks as well as single flip-flops, are thought to be organised in a single set of data words, and each data word has a unique address for identification. Hence, the information about a data modification that needs to be exchanged between redundant PESs for state restoration includes the new data value and the address. Subsequently, the main problems in state restoration at runtime are outlined.

Dependence of Computing Performance on Transfer Bandwidth

The HiQuad architecture [54] applied by HIMA is an example of a PES architecture that supports state restoration at runtime. As Figure 7.8 shows, it has a 2oo4D architecture, which means that at least *2 out of 4* redundant processing units must return equal results to maintain operation. The 'D' in the notion 2oo4D indicates that extensive diagnostics are applied in the spatially distributed nodes to detect processing errors. As typical for 2oo4 architectures, the HiQuad architecture is built of two processing units that are configured locally redundant in one rack, and two of such racks are installed spatially distributed. Most modern fault-tolerant control systems for highly safety-critical applications have such a 2oo4 or 2oo4D architecture. This is because of the lower ratio of hardware costs to failure improbability in comparison to a 2oo3 architecture. Provided multiple processing errors do not occur simultaneously, both kinds of architectures can detect any processing failure by majority voting, but a 2oo3 architecture requires spatial distribution of three racks. Figure 7.8 also shows that redundancy must be provided throughout the entire system, from the sensors *via* communication interfaces and processing units to the actors.

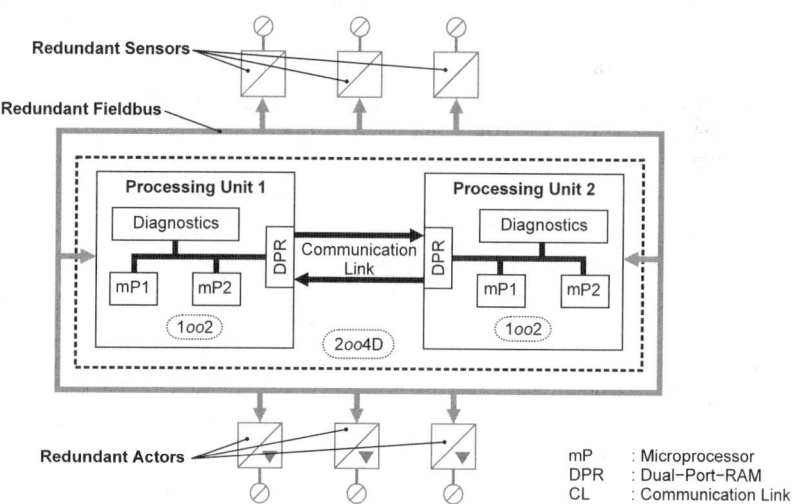

Fig. 7.8. The HiQuad-Architecture of HIMA [54]

In the HiQuad architecture, four processors are arranged in pairs of two physically separated units. Besides the two processors, each unit contains a dual-port RAM and extensive diagnostic logic to detect processing errors. The dual-port RAMs of these two units are connected *via* a communication link. If a unit is replaced, this communication link is used to copy the internal state at runtime. Obviously, the transfer bandwidth of this communication link must equal at least the frequency of write accesses to the dual-port RAMs. This points to the major problem of state restoration at runtime.

On the one hand, the redundant processing units must be installed spatially distributed in order to prevent common cause failures. As an example, to prevent a small fire simultaneously destroying all control units of a power plant, they should be installed in separate buildings. This demand for spatial distance between redundant units implies that the transfer bandwidth between them is limited. On the other hand, there is a high communication demand between the redundant units. This is because re-starting or replacing a PES unit without interrupting the on-going process requires one to equalise the unit's internal state to that of the redundant counterparts at runtime. Unfortunately, while a PES copies the internal state from its running counterparts, the latter continuously change their internal states. This aggravates the situation, since frequently changed data words might need to be transferred multiple times. Hence, the major problem associated with state restoration at runtime can be formulated as:

> *The capability for state restoration at runtime causes the attainable computing performance to be strongly dependent on the available transfer bandwidth.*

Hence, an important question is how to minimise the required transfer bandwidth.

Organising Software in Cyclically Processable Fragments

If we analyse the execution of a typical real-time application, we discover that a huge fraction of the data stored in the data memory represents intermediate values, which are irrelevant in terms of state restoration as long as the associated final data values are copied within the restoration process. Because of this, the required transfer bandwidth can be reduced through cyclic operation: intermediate values within a processing cycle are not considered for state restoration, only the final results at a cycle's end are copied. The drawback of this operating principle is that data changes can only be transferred after a processing cycle has been completed. Figure 7.9 illustrates this.

The figure shows that only a fraction of the cycle time is used for actual program execution. After program execution, the remaining cycle time is spent transferring the latest data changes. This concept transfers only RAM data modified at least once. In order to transfer RAM data that are not modified, the complete RAM content can be transferred in consecutive fragments. It is most beneficial to transfer these fragments during actual program execution. Otherwise, the transfer medium would remain unused during that period.

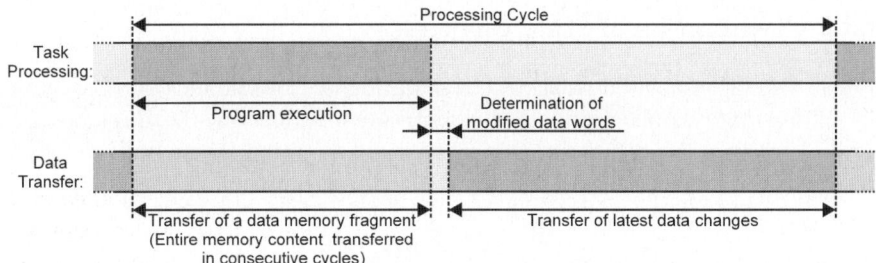

Fig. 7.9. Cyclic software execution can provide higher computing performance for a given transfer bandwidth

Of course, the restriction to cyclic software execution has a negative impact on computing performance. Processing inside each PES, however, is less restricted as the feasible transfer bandwidth. That means the computing performance inside each PES can easily be increased through a higher clock frequency, a higher data word width, or other architectural features (*e.g.*, instruction pre-fetching or pipelining). Increasing the transfer bandwidth is more costly, especially since the data exchange must suffice real-time demands.

There is another fact that makes cyclic software execution the more appropriate approach in terms of state restoration. Certainly even the best synchronisation mechanisms cannot prevent small synchronisation deviations between redundant PES instances. Additionally, the transfer of data from one PES to another always involves a delay. The use of fibre optics can decrease the actual transfer latency so that it becomes negligible, but the conventional circuitry needed for the coupling to fibre optics and for applying Error Detection/Correction Codes still causes significant delays. In the case of cyclic software execution, these synchronisation deviations and transfer delays must be taken into account only once for every cycle, *i.e.*, a new cycle cannot begin before all redundant PESs received and evaluated the transferred data of the previous cycle. A non-cyclic mode of operation requires to consider these delays in any processing step (actually, cyclic processing cannot be circumvented, since these processing steps represent an instruction *cycle* of the processor). This clearly limits the maximum clock frequency.

As major drawback, the cyclic execution principle requires one to organise software in cyclically processable fragments. Most contemporary systems, like, *e.g.*, the HiQuad system of HIMA, solve this problem by following the classical operating principle of PLCs. Nevertheless, it can be stated that:

> Cyclic software execution can provide higher computing performance
> for a given transfer bandwidth, but raises the problem of organising
> the software in cyclically processable fragments.

Obviously, a short cycle duration is desirable, since the minimum feasible response time is directly connected with it.

7.3.2 Classification of State Changes

The problems associated with state restoration at runtime have already been solved for many applications. RAID controllers, for instance, can restore the data of replaced hard disks in the background by using the periods between disk accesses for copying [91]. During the copying process, state changes of the source disks can be taken directly into account by forwarding the corresponding write accesses to the replaced disk.

Unfortunately, state restoration of real-time systems is more complex. Since they must be capable of handling several simultaneous events concurrently, the data modifications important for copying the internal state are of two different classes: *Program-controlled* and *Event-controlled Data Modifications* (*PDMs* and *EDMs*).

The PDMs comprise all data modifications the application software induces during program execution. The volume of PDMs can be bounded, *e.g.*, by restricting the number of write accesses permitted within a specified timeframe. Appropriate compiler directives are easily realisable. Of course, the restriction to a number allowing continuous transfer of information about PDMs, which comprises the new data values as well as their memory addresses, strongly limits the computing performance achievable. The latter, however, is not of major concern in terms of safety, and the high bandwidths of modern transfer technologies allow for performance more than sufficient for most highly safety-critical applications.

Limiting the volume of EDMs by an upper bound is more complex. Since digital systems always operate in discrete time steps, several events can fall within one clock cycle and, thus, occur virtually simultaneously. Moreover, a single event can cause – in comparison with the program-controlled situation – a relatively large number of data modifications within an extremely short period of time. For example, in principle it is possible that a single event activates all tasks of an application at once. In this case, storing the activation time for each task would result in a huge amount of EDMs. Obviously, restricting the frequency of write accesses would have a strong negative effect on real-time performance, *e.g.*, it would limit the minimum realisable response time. Thus, the capability to transfer all information about EDMs at runtime necessitates a trade-off between restricting the frequency of EDMs to a volume that allows for acceptable real-time performance and the transfer bandwidth required. To sum up:

> *Restricting the frequency of PDMs affects only the computing performance, whereas restricting the frequency of EDMs has a negative impact on the real-time performance.*

That is why in task-oriented real-time PESs, which must provide both practicable computing performance and very short response times, sophisticated methods need to be applied to enable state restoration at runtime.

7.3.3 State Restoration with Modification Bits

Section 7.3.1 showed why it is beneficial to combine the capability for state restoration at runtime with a cyclic operating principle. This allows a higher computing performance for a given transfer bandwidth. Hence, the concept for task-oriented real-time execution introduced in the previous chapter, which is based on a cyclic operating principle, is perfectly suited for state restoration at runtime.

Obviously, a short cycle duration is desirable, since the shorter the cycle time the shorter is the minimum realisable response time. If the information about a cycle's data modifications is transferred immediately, *i.e.*, the associated data values and addresses are transferred at the cycle's end, there are two facts that render minimising the cycle duration problematic:

- Immediate transfer of a cycle's *Program-controlled Data Modifications (PDMs)* requires one to limit the number of PDMs that Execution Blocks are allowed to perform within a processing cycle. Appropriate directives are easy to implement in the compiler software that fragments application programs into Execution Blocks. The PDM limit has, however, a negative impact on computing performance, since it represents an additional restriction for the size of program code fragments.
- Unfortunately, the minimum number of *Event-controlled Data Modifications (EDMs)* is independent of the cycle duration, *i.e.*, the minimum number of data modifications that can occur within a cycle and that must be considered for state restoration does not depend on the cycle length. This number results from a worst-case scenario, *e.g.*, all tasks are activated within the same cycle, causing several new activation time values at once. The capability for immediate transfer of this number of EDMs necessitates a certain amount of data bits to be transferred any cycle. Consequently, the computing efficiency, *i.e.*, the ratio of time spent for actual program execution and time spent for data transfer, decreases with decreasing cycle durations.

These problems are prevented by the use of *Modification Bits*, which is the underlying principle of the approach to state restoration proposed in [10]. Every data word is assigned a *Modification Bit*, which signals whether the data word has been modified. This bit is set when the data word's value is changed; it is reset when the new value has been transferred to the redundant PESs. Instead of transferring information about data modifications immediately, *i.e.*, within the cycle they occurred in, the values and addresses of a fixed amount of data words are transferred at the end of any cycle. Obviously, the transfer includes only those data words whose Modification Bits are set.

This technique has the advantage that the transfer bandwidth available does not need to suffice the maximum number of data modifications within a cycle. In other words, the approach does not necessitate exchange at any cycle's end as many data bits as the transfer of the information about a cycle's data modifications would require in the worst case. Instead, the average

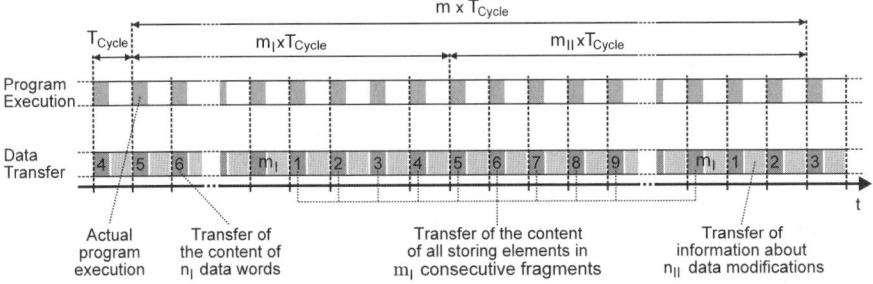

Fig. 7.10. The state restoration process consists of two phases: phase I ensures that the content of all storage elements is copied at least once; phase II endures until all recent data modifications have been transferred at a cycle's end

frequency of data modifications defines the amount of data that must be exchanged any cycle to enable state restoration. As a drawback, the technique increases the effort to determine the maximum duration of the state restoration process, or – what means the same, but has practical benefits – to prove that state restoration will complete within a given time frame. This is because it is necessary to know for every data word its maximum number of value changes within a given time frame. This is not easy to determine for real-time systems that follow the approach of asynchronous programming.

If the maximum frequencies of value changes are known, the amount of data to be transferred any cycle is computable. The calculation depends on how data words are handled whose values are not changed within the specified time frame. One approach would be to set the Modification Bits of all data words at the begin of the restoration process. Thus, they will be copied at least once. In this approach, the transfer medium remains unused during actual program execution, since the information on modified data is only available after it. That is why better performance is provided by the policy depicted in Figure 7.9, which uses the transfer medium during actual program execution to transfer the content of all storage elements in consecutive fragments. This approach does not add the handling of non-modified data words to the transfer after actual program execution and, thus, requires less data to be restored by the Modification Bit scheme.

In order to prove that state restoration completes within m cycles, the state restoration process is divided into two phases. Figure 7.10 illustrates the subsequent description:

Phase I. The first phase ensures that the content of all storage elements is copied at least once. If α is the number of data words into which the storage elements are organised, and n_I is the number of data words whose content is transferred during actual program execution of a cycle, then phase I endures

$$m_I = \left\lceil \frac{\alpha}{n_I} \right\rceil \tag{7.2}$$

cycles. Thus, after m_I consecutive cycles, all data words have been copied at least once, and the Modification Bit scheme needs to keep track of data modifications, only.

Phase II. The second phase, which endures $m_{II} = (m - m_I)$ cycles, keeps track of on-going data modifications until state restoration completes. For this, there must be at least one cycle within phase II after which all data modifications since the begin of phase I have been transferred.

This is the case if the number of data modifications whose information is transferred at any cycle's end equals

$$n_{II} \geq \frac{1}{m_{II}} \times (\alpha + \beta) \quad , \tag{7.3}$$

with

$$\beta = \sum_{i=1}^{\alpha} \left\lfloor \frac{m_{II} \times T_{Cycle}}{P_i} \right\rfloor \quad . \tag{7.4}$$

In the latter equation, P_i is the minimum period between two value changes of the i-th data word. The formula results from the fact that every data word i is changed at most one time more than period P_i completely fits in m_{II} cycles. The quantity n_{II} does not necessarily need to be an integer number. Since it is possible to transfer the information about a data modification within a number of consecutive cycles, n_{II} can be any rational number.

Equation 7.3 allows one to illustrate the reduction of the bandwidth requirements in an example. Consider a set of three data words with the value change periods $p_1 = 2 \times T_{Cycle}$, $p_2 = 3 \times T_{Cycle}$, and $p_3 = 6 \times T_{Cycle}$. The number of cycles granted for state restoration be 100. In the case of immediate transfer, information about three data modifications would need to be transferred at each cycle's end, since all three data words could change their values simultaneously. Applying Equation 7.3 yields that information about only 1.02 data modifications needs to be transferred at the end of each cycle.

The major drawback of this state restoration policy is that the handling of the Modification Bits causes additional computing effort. This makes a software-based implementation as presented in [10] inefficient. In particular, the search for data words with set Modification Bits strongly restricts the achievable performance, since – in the worst case – all Modification Bits must be evaluated. This problem can be solved by implementing the TAU in form of a dedicated logic circuitry. As shown in [101], this allows one to identify the data words with set Modification Bits in parallel to program code execution.

7.3.4 Concept of State Restoration

The concept of task-oriented real-time execution without asynchronous interrupts described above enables a straightforward implementation of the modification bit scheme for state restoration at runtime. Since it already founds on a cyclic operating principle, it is just necessary to add the administration of *Modification Bits*, and to lengthen the cycle duration for the transfer of a fixed amount of modified data words. The strict separation of task administration and task execution implies that the associated data are also handled separately:

- The *Task Administration Unit's (TAU)* data relevant for state restoration comprise the Task List Memory (TLM) and the ID of the Execution Block to be executed in the subsequent cycle. The latter consists of a few bits, only, and is transferred at the end of any cycle. Hence, only the TLM needs to be restored following the modification bit scheme.

  For this, each data word of the TLM is assigned a Modification Bit, and a dedicated digital circuitry, which is described in [101], administrates these Modification Bits during Sequential Task Administration (STA). The circuit enables access to the modified data words directly after completion of STA, *i.e.*, it is not necessary to search for data words with set Modification Bits.

- Since the content of the processor registers is lost at the end of any Execution Cycle, only the data memory of the Application Processing Unit (APU) is relevant for state restoration. As for the TLM of the TAU, to every data word of the memory a Modification Bit is assigned. A dedicated circuitry, which is described in [101], administrates these Modification Bits and enables access to the modified data words directly after completion of an Execution Block, *i.e.*, there is no additional delay for the determination of modified data words.

The immediate access to data words with set Modification Bits enables the transfer scheme described in the sequel.

Transfer Schedule for Serial Data Streams

As described at the begin of this section, the state restoration data are transferred *via* Serial Data Streams (SDSs), which are exchanged between redundant PES instances. Figure 7.11 shows the transfer schedule of these SDSs. The SDSs' Transfer Cycles correspond to the PESs' Execution Cycles, except that they are shorter by a small idle gap. At the beginning of any Transfer Cycle, the SDSs transfer information about asynchronous events detected during the previous cycle. This does not serve actual state restoration, but is necessary to ensure identical processing of redundant instances. Due to signal delays, asynchronous events may not be detected simultaneously by all redundant units. Additionally, there might be slight synchronisation deviations between their internal clocks. Hence, an asynchronous event that occurs shortly before a cycle's end might not be detected by all PESs in the current cycle; some PESs may detect it in the subsequent one. The event information transferred by the SDSs enables one to shift an event signal to the next cycle if it has not been detected by all redundant units. After the event information, the SDSs transfer a fragment of the data contained in the TLM and the APU data memory. This is done in a way that the entire content of the TLM and the APU data memory is transferred in m_I consecutive cycles.

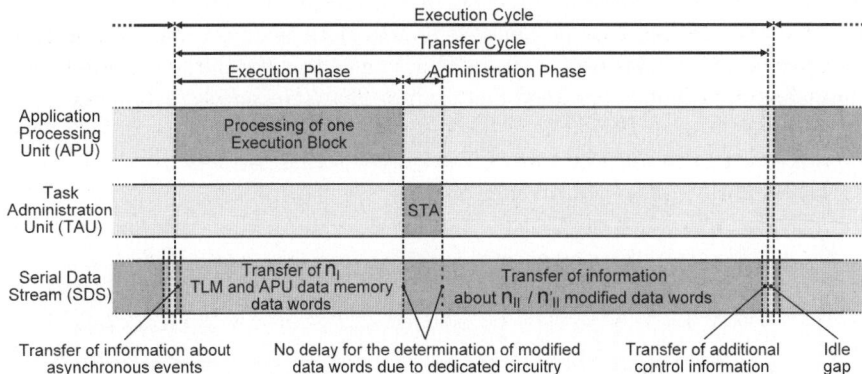

Fig. 7.11. Transfer schedule of *Serial Data Streams* in relation to Execution Cycles

Immediately after the APU has completed an Execution Block, the SDSs start to transfer information about n_{II} data modifications. Since this transfer starts before task administration finishes, at first only recent changes in the APU data memory are handled. Information about modifications of the TLM are transferred after task administration has been completed. The fact that at most n'_{II} data modifications in the TLM are transferred, any cycle is compensated by the fact that TLM modifications are handled with higher priority than modifications of APU data memory. This means that the latter are only transferred if no TLM modifications need to be transferred, *i.e.*, all Modification Bits associated to data words of the TLM are in the reset state.

At the end of any transfer cycle, additional control information is transferred *via* SDSs. This comprises the ID of the Execution Block to be executed in the next cycle as well as the binary information whether all recent data modifications have been transferred. A restarted PES uses this binary information to identify the end of the state restoration process. Figure 7.11 also shows a small gap between a transfer cycle's end and the beginning of the next cycle. This idle gap is necessary to compensate for small synchronisation deviations between redundant PES instances. The gap ensures that the PES, whose internal clock runs ahead most, does not start a new Execution Cycle before it completely received the SDS of the PES, whose clock lags behind most.

State Restoration Process

In order to verify the correctness of received SDSs by majority voting, not only the internal task processing of all redundant PESs must be identical, but also the state restoration data transferred. This is why states are restored in synchrony with the global time standard Universal Time Co-ordinated (UTC) at certain UTC-synchronous instants, which coincide with a cycle's beginning, all Modification Bits are reset. Thus, assuming that the internal task processing – which is also in reference to UTC – of all redundant PESs is identical, the modification bits will also be handled in an identical way from that moment on. These UTC-synchronous instants which, *e.g.*, can match the begin of UTC minutes, are subsequently called *Restoration Synchronisation Instants (RSIs)*. Obviously, state restoration must be completed between two such RSIs. There are always m cycles between two RSIs. Figure 7.12 illustrates the subsequent description of the state restoration process.

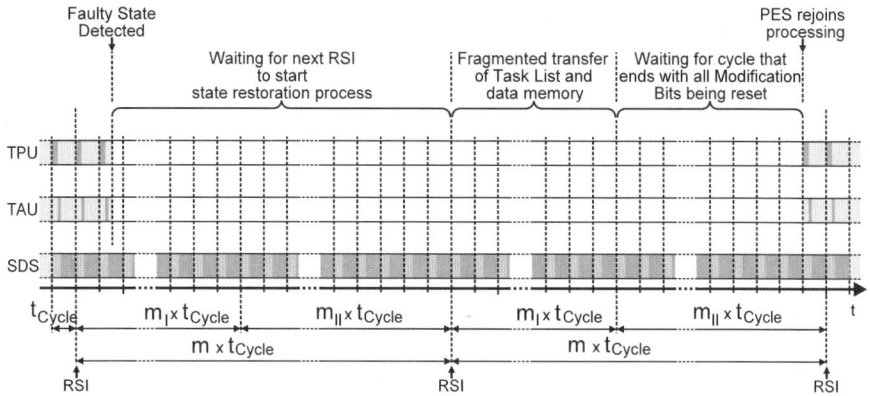

Fig. 7.12. State restoration process

Assume a PES assesses its state to be faulty, *e.g.*, because its own SDS deviates from the other SDSs' majority. In this case, the PES re-starts itself and initiates state restoration. For this, it must wait for the next RSI, before the state restoration process can begin. In the worst case, the last RSI has just occurred, and the PES must wait for m cycles, where $m \times T_{Cycle}$ is the period between two RSIs.

At the moment of the next RSI, which coincides with a cycle's begin, all Modification Bits are reset and the state restoration process begins. During the subsequent m_I cycles, the entire content of the TLM and the APU data memory is transferred in consecutive fragments. Even if one of these m_I cycles ends with all Modification Bits being in reset state, *i.e.*, all recent data modifications of the running PESs have been copied, the re-started PES cannot re-join the redundant processing at the subsequent cycle's beginning. This is because – so far – some data words might have neither been covered by the fragmented transfer in consecutive cycles nor by the Modification Bit scheme.

When the last of the m_I cycles has elapsed, all data words of the TLM and the APU data memory have been copied to the re-started PES at least once.

Furthermore, the Modification Bit scheme kept track of all data modifications that happened since the last RSI. Hence, during the remaining m_{II} cycles before the next RSI, the PES just needs to wait for a cycle that ends with all Modification Bits being in reset state, *i.e.*, with all recent data modifications of the running PESs being copied. As soon as such a cycle occurs, the state restoration is complete, and the re-started PES can re-join the redundant processing in the subsequent cycle. As will be explained below, the fulfillment of two conditions similar to Equation 7.3 ensures that state restoration completes within these m_{II} cycles in any case.

Since a re-started PES must wait m cycles in the worst case before the state restoration process begins, and this process endures at most m cycles, the worst-case duration of state restoration is $2{\times}m$ cycles. In other words, the time elapsing from the detection of a PES's faulty state until the end of the restoration process is at most

$$t_{StateRestoration} = 2 \times m \times T_{Cycle} \ . \tag{7.5}$$

7.3.5 Influence on Program Code Fragmentation and Performance Aspects

Obviously, the capability for state restoration at runtime causes the frequency of data modifications to be another feasibility criterion. Appropriate feasibility conditions have been derived and can be found in [101]. Besides the real-time check for temporal feasibility, it is necessary to verify that the state restoration process completes even in the worst-case scenario. This arises when all tasks simultaneously execute the program paths inducing the highest numbers of data modifications.

As a consequence, there are two dimensions in optimising program code fragmentation. The first focuses on the number of Execution Blocks that must be processed on the longest program path of a task. Since this number corresponds to the WCET, its minimisation is an optimisation criterion aiming for temporal feasibility (*cf.* Section 7.2.4). The second dimension focuses on the worst-case number of data modifications, *i.e.*, the number that arises on the program path with the highest number of data modifications. Reducing the amount of data that must be passed from one Execution Block to the next is an optimisation criterion that aims for feasibility of the state restoration process.

The program path causing the WCET is not necessarily the same as the one that gives rise to the highest number of data modifications. Nevertheless, the two optimisation dimensions are not contradictory. In fact, the associated optimisation strategies are very similar. On the fragmentation level, the WCET is implicitly minimised by aiming for high block utilisation, *i.e.*, a low percentage of idle processing. As discussed in Section 7.2.4, a high block utilisation can be achieved by starting from the finest possible block granularity followed by an optimised merging of code fragments. Instead of solely minimising idle processing, this merging can also be carried out with the objective of processing intermediate values in as few blocks as possible. This lowers the number of intermediate values that must be exchanged between Execution Blocks *via* the data memory and, hence, reduces the number of data modifications. Nevertheless, there are scenarios in which improvements in one dimension negatively affect the other.

Obviously, the computing performance achievable depends strongly on the quality of the code fragmentation algorithm. This makes it difficult to discuss performance aspects without referring to a sophisticated algorithm. A practical approach is to discuss computing performance in comparison with existing systems.

- **Comparison with the HiQuad-Architecture of HIMA**

  The programmable electronic systems H41q/H51, which are manufactured by HIMA, use the HiQuad architecture described in Section 7.3.1 to support state restoration at runtime. They also operate in a cyclic fashion, and are programmed in accordance with the programming paradigm defined by IEC 61131. Unfortunately, the system descriptions provided by HIMA do not explain the state restoration strategy applied. Of course, if they had implemented a sophisticated restoration scheme, they would probably not publish its details. Even a patent protection, which would require a detailed publication, might not prevent competitors from copying, since it is difficult to prove that an integrated circuit internally applies a certain technique. Nevertheless, if HIMA applied a certain technique, we assume that they would at least mention it in their advertising brochures.

  That is why we assume that the data modifications of a cycle are simply transferred at its end. Probably, implementing a special state restoration technique has been considered unnecessary, since it only increases performance, and performance is not the major design criterion for such systems. In other words, the associated performance gain does not justify the higher effort for safety licensing. The systems follow the approach of synchronous programming, and execute the same program code in any cycle. Consequently, nearly the same data words are modified in any cycle. This makes the Modification Bit technique inefficient.

  The synchronous programming style is, however, very inflexible and its field of application is limited to simple control tasks. Thus, the performance advantage of the PES concept described here is the higher flexibility that task-oriented real-time execution in discrete cycles provides.

- **Comparison with the approach of [10]**

  The approach presented in [10] is also based on Modification Bits. The major difference to the PES concept shown here is that the Modification Bits are administrated by software – after actual program execution. This causes a significant delay, since – in the worst case – the entire data memory must be searched for set Modification Bits.

  The described concept prevents this delay through dedicated circuitry. Unlike the software-based solution, this circuitry allows one to use the transfer medium during actual program execution for a fragmented transfer of the entire data memory. Thus, the Modification Bit scheme must only cover the data words modified during the restoration process; unchanged data words are covered by the fragmented transfer. To realise this efficiently, all Modification Bits are reset at UTC-synchronous instants, whereas the concept of [10] sets all Modification Bits whenever a processing error is detected. Obviously this reduces the amount of data that must be transferred after actual program execution and, hence, higher computing performance is achievable for a given transfer bandwidth. The UTC-synchronisation has the additional advantage that the redundant systems need no further synchronisation to proceed with state restoration.

In summary, the discussion showed that the described PES concept – in particular its state restoration technique – has valuable performance advantages over existing state restoration concepts. Further information about the concept can be found in [101].

8

Epilogue

Pervading areas and carrying out control functions which were only recently unthinkable, embedded programmable electronic systems have found their way into safety-critical applications. In the general public, however, awareness is rising of the inherent safety problems associated with computerised systems, and particularly with software. Taking the high and fast increasing complexity of control software into account, it is obvious that the problem of software dependability will likewise multiply.

There has always existed a mismatch between the design objectives for generic universal computing on one hand and for embedded control systems on the other. Practically all sophisticated dynamic and "virtual" features devised in the so-called "high technology" world of computers to match obsolete approaches from the past and aiming to enhance the average performance of computers must be considered harmful for embedded systems. Thus, inappropriate categories, such as probabilistic and statistical terms or fairness, and optimality criteria, such as minimisation of average reaction time, must be replaced by recognising the constraints imposed by the real world, *i.e.*, by the notion of resource adequacy.

Embedded systems have to meet timing conditions. Although guidelines for proper design and implementation of embedded control systems operating in real-time environments have been known for a long time, in practice *ad hoc* approaches still prevail to a large extent. This is due to the fact that the notion of time has long been — and is still mostly being — ignored as a category in computer science. There, time is reduced to predecessor-successor relations, and is abstracted away even in parallel systems. In standard programming environments, no absolute time specifications are possible, the timing of actions is left implicit, and there are no time-based synchronisation schemes.

The prevailing methods and techniques for assessing embedded systems are based on testing, and the assessment quality achieved with them mainly depends on the designers' experience and intuition. It is almost never proven at design time that such a system will meet its temporal requirements in any situation that it may encounter. Although this situation was identified several decades ago, it has not been improved because there are no modern and powerful processors with easily predictable behaviour, nor compilers for languages that would prevent writing software with unpredictable runtimes. As a result of all of this, commercial off-the-shelf control computers are generally not suitable for safety-critical applications.

Against the background outlined above, it is the objective of this book to promote adequate and consistent design of embedded control systems with dependability and, particularly, safety requirements, of the least demanding safety integrity level SIL 1, by presenting contributions aiming to improve both functional and temporal correctness in a holistic manner on the basis of a unified concept for safety functions. It is the aim to reach the state where computer-based systems can be constructed with a sufficient degree of confidence in their dependability. To this end, semantic gaps are to be prevented from arising, difficulties are to be prevented by design instead of handling them upon occurrence, and strict specification and verification are to be integrated into the design process in a problem-oriented way, without imposing too much additional effort on often already overloaded application designers.

In striving to meet this objective, certain peculiarities of embedded systems need to be observed, namely that it is often necessary to develop not only their software but also their hardware, and sometimes even their operating systems, and that optimum processor utilisation is not so relevant for them, as costs have to be seen in the framework of the external processes controlled, and with regard to the latters' safety requirements. Further, instead of increasing processing power, technological advances should be utilised to free chip space for accommodating application-oriented on-chip resources. It has also to be kept in mind that developers need to convince official bodies that they have identified and dealt with all relevant hazards, as safety of control systems needs to be established by certification.

Towards eliminating the shortcomings of current practice and achieving the objectives mentioned above, the following contributions were made in this book.

Predictability of temporal behaviour was identified as the ultimate property of embedded real-time systems. It was suggested that one base this on a comprehensive utilisation of an adequate notion of time, *viz.*, Universal Time Co-ordinated, for providing programming language support for temporal predictability and, if realistic determination of execution time by source code analysis is a goal, for devising simpler processor architectures.

Actually, simplicity is a means to realise dependability, which is the fundamental requirement for safety-related systems. Simplicity turned out to be a design principle fundamental to fight complexity and to create confidence. Design simplicity prevents engineering errors and, later, eases safety licensing. It is much more appropriate to find simple solutions, which are transparent and understandable and, thus, inherently safer. Such adequate solutions are characterised by simple, inherently safe programming, are best on the specification level, re-use already licensed application-oriented modules, use graphics instead of text, and rigorous — but not necessarily formal — verification methods understandable by non-experts such as judges. The more safety-critical a function, the simpler the related software and its verification ought to be. It was advocated to use adequate programming methods, languages and tools as well as problem-oriented instead of primitive implementation-oriented scheduling methods.

Designers tend to deal with unwanted events as exceptions. It is, however, irrelevant whether they are unwanted or not. If they can be anticipated during the design phase, they should be included in the specifications to be handled adequately. To this end, the aspects of reconfiguring computer control systems with special emphasis on the support of methods for higher-level control system reconfiguration, and of recovery were considered in detail. With respect to the latter, only forward recovery is possible for real-time systems, to bring them into certain predefined, safe, and stable states.

To ease the provision of fault tolerance, a case was made for distributed asymmetrical multiprocessor architectures with dedicated processors. It was shown how to solve problems with appropriate approaches, *e.g.*, jitter was fully eliminated by hardware support. Moreover, it was shown that the schedulability of tasks can always be ensured by employing feasible scheduling policies. Traditional elements of real-time systems were represented as objects, thus introducing object orientation to the design of embedded systems.

In the previous chapter, a consistent architectural concept for safety-related programmable electronic systems based on a novel and patented operation paradigm was presented, which combines the benefits, but eliminates the drawbacks of synchronous and asynchronous programming. It features task-oriented real-time execution without the need for asynchronous interrupts, and a high degree of fault tolerance by the ability for state restoration at runtime from redundant system instances as a form of forward recovery, thus enabling one to employ control software structured in the form of tasks even for applications having to meet the requirements of safety integrity level SIL 3.

References

1. Ada83 Language Reference Manual,
   http://www.adahome.com/lrm/83/rm/rm83html/, 1983.
2. Ada95 Reference Manual, http://www.adahome.com/rm95/, 1995.
3. ANSI/ISA S84.01: Application of safety instrumented systems for the process industry. American National Standards Institute, 1996.
4. John Backus. Specifications for the IBM Mathematical FORmula TRANslating system, FORTRAN. Technical report, IBM Corporation, New York, NY: IBM Corporation, 1954.
5. R. Belchner. Flexray requirements specification (draft). Version 1.9.7. http://flexray-group.com, 2001.
6. M. Ben-Ari. *Principles of Concurrent Programming*. Prentice-Hall, 1982.
7. Guillem Bernat, Antoine Colin, and Stefan M. Petters. WCET analysis of probabilistic hard real-time systems. In *RTSS, Real-Time Systems Symposium*, Austin, TX, USA, December 2002. IEEE.
8. Enrico Bini, Giorgio C. Buttazzo, and Giuseppe M. Buttazzo. Rate monotonic analysis: The hyperbolic bound. *IEEE Transactions on Computers*, 52(7):933–924, July 2003.
9. Andrew P. Black. Exception handling: The case against. Technical Report Technical Report TR 82-01-02, Department Of Computer Science, University of Washington, May 1983. (originally submitted as a PhD thesis, University of Oxford, January 1982).
10. A. Bondavalli, F. Di Giandomenico, F. Grandoni, D. Powell, and C. Rabéjac. State restoration in a COTS-based N-modular architecture. In *1st IEEE Int. Symposium on Object-oriented Real-time distributed Computing (ISORC '98)*, pages 174–183, Kyoto, Japan, April 20 - 22 1998.
11. Alan Burns and Andy Wellings. *Real-Time Systems and Programming Languages. Second Edition*. Addison-Wesley Publishing Company, 1996.
12. Giorgio C. Buttazzo. Rate monotonic vs. EDF: Judgment day. *Real-Time Systems*, 29(1):5–26, January 2005.
13. Anton Cervin. *Integrated Control and Real-Time Scheduling. ISRN LUTFD2/TFRT-1065-SE*. PhD thesis, Department of Automatic Control, Lund University, Sweden, 2003.
14. K.-M. Cheung, M. Belongie, and K. Tong. End-to-end system consideration of the Galileo image compression system. Technical report, Telecommunications and Data Acquisition Progress Report, April–June 1996.

15. W.J. Cody, J.T. Coonen, D.M. Gay, K. Hanson, D. Hough, W. Kahan, J. Palmer R. Karpinski, F.N. Bis, and D. Stevenson. A proposed radix- and word-length-independent standard for floating-point arithmetic. *IEEE Micro*, 4(4):86–100, August 1984.

16. Matjaž Colnarič. *Predictability of Temporal Behaviour of Hard Real-Time Systems*. PhD thesis, University of Maribor, June 1992.

17. Matjaž Colnarič and Wolfgang A. Halang. Architectural support for predictability in hard real-time systems. *Control Engineering Practice*, 1(1):51–59, February 1993. ISSN 0967–0661.

18. Matjaž Colnarič, Domen Verber, Roman Gumzej, and Wolfgang A. Halang. Implementation of embedded hard real-time systems. *International Journal of Real-Time Systems*, 14(3):293–310, May 1998.

19. Matjaž Colnarič, Domen Verber, and Wolfgang A. Halang. A Real-Time Programming Language as a Means to Express Specifications. In *WRTP'96, Proceedings of IFAC Workshop on Real-Time Programming*, Gramado, RS, Brazil, November 1996.

20. SystemC Community. OSCI SystemC TLM 2.0 – Draft 1. http://www.systemc.org/web/sitedocs/tlm_2_0.html, February 2007.

21. James E. Cooling. *Software Engineering for Real-Time Systems*. Addison Wesley, 2003.

22. James E. Cooling and P. Tweedale. Task scheduler co-processor for hard real-time systems. *Microprocessors and Microsystems*, 20(9):553–566, 1997.

23. Flaviu Cristian. Exception handling and software fault tolerance. *IEEE Transactions on Computers*, 31(6):531–540, June 1982.

24. Flaviu Cristian. Correct and robust programs. *IEEE Transactions on Software Engineering*, 10(2):163–174, March 1984.

25. L. F. Currie. *High-integrity Software*, chapter Newspeak - a reliable programming language, pages 122–158. Computer Systems Series. Pitman, 1989.

26. Michael L. Dertouzos. Control robotics: The procedural control of physical processes. In *Proceedings of IFIP Congress*, pages 807–813, 1974.

27. DIN 44 300 A2: Informationsverarbeitung. Berlin–Cologne, 1972.

28. DIN 66 201 Part 1: Prozessrechensysteme. Berlin–Cologne, 1981.

29. DIN 66 253: Programming language PEARL, Part 1: Basic PEARL. Berlin, 1981.

30. DIN 66 253: Programming language PEARL, Part 3: PEARL for distributed systems. Berlin, 1989.

31. DIN V VDE 0801: Grundsätze für Rechner in Systemen mit Sicherheitsaufgaben (Principles for using computers in safety-related systems). VDE Verlag, 1990.

32. DIN V 19250: Leittechnik - Grundlegende Sicherheitsbetrachtungen für MSR-Schutzeinrichtungen (Control technology; Fundamental safety aspects for measurement and control equipment). VDE Verlag, 1994.

33. Embedded C++ Technical Committee. *The Embedded C++ specification*. http://www.caravan.net/ec2plus/language.html, October 1999.

34. EN 50126: Railway applications – The specification and demonstration of dependability, reliability, availability, maintainability and safety (RAMS). Comité Européen de Nomalisation Electrotechnique, 1999.

35. EN 50128: Railway applications – Software for railway control and protection systems. Comité Européen de Nomalisation Electrotechnique, 2002.

36. EN 50129: Railway applications – Safety related electronic systems for signalling. Comité Européen de Nomalisation Electrotechnique, 2002.
37. EN 954: Safety of machinery – Safety-related parts of control systems. Comité Européen de Nomalisation Electrotechnique, 1996.
38. EUROCAE-ED-12B: Software considerations in airborne systems and equipment certification. (european equivalent to the US standard RTCA DO-178B). European Organisation for Civil Aviation Equipment., 1992.
39. Max Felser. Real-time Ethernet – industry prospective. *Proceedings of the IEEE*, 93(6), June 2006.
40. J. Fonseca, F. Coutinho, and J. Barreiros. Scheduling for a TTCAN network with a stochastic optimization algorithm. In *Proceedings 8th International CAN Conference*, Las Vegas, Nv., 2002.
41. T. Fredrickson. Relationship of EN 954-1 and IEC 61508 standards. Safety Users Group., 2002.
42. Alceu H. Frigeri, Carlos E. Pereira, and Wolfgang A. Halang. An object-oriented extension to PEARL 90. In *Proc. 1st IEEE International Symposium on Object-Oriented Real-Time Distributed Computing, Kyoto*, pages 265–274, Los Alamitos, 1998. IEEE Computer Society Press.
43. Thomas Führer, Bernd Müller, Werner Dieterle, Florian Hartwich, Robert Hugel, and Michael Walther. Time triggered communication on CAN. Technical report, Robert Bosch GmbH, http://www.can.bosch.com/, 2000.
44. Mohammed G. Gouda, Yi-Wu Han, E. Douglas Jensen, Wesley D. Johnson, and Richard Y. Kain. Distributed data processing technology, Vol. IV, Applications of DDP technology to BMD: Architectures and algorithms, Chapter 3, Radar scheduling: Section 1, The scheduling problem. Technical report, Honeywell Systems and Research Center, Minneapolis, MN, Sepember 1977.
45. TTA Group. Time-triggered protocol (TTP/C), version 1.0. http://www.ttagroup.org/ttp/specification.htm, 2001.
46. Jan Gustafsson. *Analyzing Execution-Time of Object-Oriented Programs Using Abstract Interpretation*. PhD thesis, Mälardalen University, May 2000.
47. Wolfgang A. Halang. Parallel administration of events in real-time systems. *Microprocessing and Microprogramming*, 24:687–692, 1988.
48. Wolfgang A. Halang and Alceu H. Frigeri. Methods and languages for safety related real-time programming. Technical report, Fernuniversität Hagen, Report on research project F 1636 funded by Bundesanstalt für Arbeitschutz und Arbeitsmedizin, Dortmund, Germany, 1999.
49. Wolfgang A. Halang and Alexander D. Stoyenko. Comparative evaluation of high level real time programming languages. *Real-Time Systems*, 2(4):365–382, December 1990.
50. Wolfgang A. Halang and Alexander D. Stoyenko. *Constructing Predictable Real-Time Systems*. Kluwer Academic Publishers, Boston–Dordrecht–London, 1991.
51. Wolfgang A. Halang and Janusz Zalewski. Programming languages for use in safety-related applications. *Annual Reviews in Control, Elsevier*, 27(1), 2003.
52. Les Hatton. *Safer C: Developing for High-Integrity and Safety-Critical Systems*. McGraw-Hill, 1995.
53. Philippe Hilsenkopf. Nuclear power plant I&C and dependability issues. Invited talk. In *Proceedings of WRTP97*, Lyon, France, 1997. IFAC.
54. HIMA. Product information: H41q/H51q safety systems. www.hima.com, 2006.

55. Hoai Hoang, Magnus Jonsson, Ulrik Hagström, and Anders Kallerdahl. Switched real-time ethernet with earliest deadline first scheduling - protocols and traffic handling. In *Proceedings of the International Parallel and Distributed Processing Symposium (IPDPS 2002)*. IEEE Computer Society, 2002.

56. IEC 1131-3: Programmable controllers, part 3: Programming languages. International Electrotechnical Commission, Geneva, 1992.

57. IEC 60880: Software for computers in the safety systems of nuclear power stations. International Electrotechnical Commission., 1987.

58. IEC 61508: Functional safety of electrical/electronic programmable electronic systems: Generic aspects. part 1: General requirements. International Electrotechnical Commission, Geneva, 1992.

59. IEC 61511: Functional safety: Safety instrumented systems for the process industry sector. International Electrotechnical Commission, Geneva, 2003.

60. IEC 61513: Nuclear power plants – instrumentation and control for systems important to safety – general requirements for systems. International Electrotechnical Commission, Geneva, 2002.

61. 1149.1 IEEE. *Test Access Port & Boundary Scan Architecture*. IEEE, New York, 1990.

62. IFATIS. Intelligent Fault Tolerant Control in Integrated Systems. IST-2001-32122; http://ifatis.uni-duisburg.de/, 2002-2004.

63. Texas Instruments. C6711 DSP Starter Kit (DSK). http://focus.ti.com/docs/toolsw/folders/print/tmds320006711.html, 2001.

64. ISO/CD 11898-4: Road vehicles - controller area network (CAN) - part 4: Time triggered communication. ISO/TC 22/SC 3/WG 1/TF 6, 2004.

65. ISO/IEC/ANSI 8652:1995: Information Technology – Programming Languages – Ada. International Electrotechnical Commission, Geneva, 1995.

66. Farnam Jahanian and Aloysius Ka-Lau Mok. Safety analysis of timing properties in real-time systems. *IEEE Transactions on Software Engineering*, 12(9):890–904, September 1986.

67. E. Douglas Jensen. Real-time for the real world. section: Time/utility functions. http://www.real-time.org/timeutilityfunctions.htm, 2005.

68. JOVIAL Program Office, http://www.jovial.hill.af.mil/. *JOVIAL Support*, 2006.

69. Eugene Kligerman and Alexander D. Stoyenko. Real-time euclid: a language for reliable real-time systems. *IEEE Trans. Softw. Eng.*, 12(9):941–949, 1986.

70. Wilfried Kneis. Draft standard industrial real-time FORTRAN. *ACM SIG-PLAN Notices*, 16(7):45–60, October 1981.

71. Hermann Kopetz. Time-triggered versus event-triggered systems. In *Proc. International Workshop on Operating Systems in the 90s and Beyond*, volume 563 of *Lecture Notes in Computer Science*, pages 87–101, Berlin, 1992. Springer Verlag.

72. Hermann Kopetz, A. Damm, Ch. Koza, M. Mulazzani, W. Schwabl, Ch. Senft, and R. Zainlinger. Distributed fault-tolerant real-time systems: The MARS approach. *IEEE Micro*, 9(1):25–40, February 1989.

73. H. Krebs and U. Haspel. Ein Verfahren zur Software-Verifikation. *Regelungstechnische Praxis rtp*, 26:73 – 78, 1984.

74. J. Labetoulle. Real-time scheduling in a multiprocessor environment. technical report. Technical report, IRIA Laboria, Rocquencourt, 1976.

75. Jean J. Labrosse. *Microc/OS-II: The Real-Time Kernel*. CMP Books, Berkeley, 1998.

76. Harold W. Lawson. Cy-Clone: An approach to the engineering of resource adequate cyclic real-time systems. *Real-Time Systems, Kluwer Academic Publishers*, 4(1):55–83, 1992.

77. Harold W. Lawson. Systems engineering of a successful train control system. Invited talk. In *Proceedings of WRTP2000*, Mallorca, Spain, 2000. IFAC.

78. David Liddell. Simple design makes reliable computers. In Michel Banâtre and Peter A. Lee, editors, *Hardware and Software Architectures for Fault Tolerance*, pages 91–94, no address, 1993. Springer.

79. Lennart Lindh. *Utilization of Hardware Parallelism in Realizing Real-Time Kernels*. PhD thesis, Royal Institute of Technology, Sweden, 1989.

80. C.L. Liu and J.W. Layland. Scheduling algorithms for multiprogramming in a hard real-time environment. *Journal of the ACM*, 20(1):46–61, 1973.

81. Mike J. Livesey and Colin Allison. A dynamically configurable co-processor for microkernels. In *EUROMICRO '94 - Workshop on Parallel and Distributed Processing*, Malaga, Spain, 1994.

82. Uwe Maier and Matjaž Colnarič. Some basic ideas for intelligent fault tolerant control systems design. In *Proceedings of 15th IFAC World Congress*, Barcelona, Spain, July 2002.

83. Microchip. *8-bit PIC® Microcontrollers*. http://www.microchip.com/, 2006.

84. MISRA-C. Guidelines for the use of the c language in critical systems. http://www.misra.org.uk, October 2004.

85. Modula-2 reference. http://www.modula2.org/reference/index.php, 2007.

86. Aloysius K. Mok, P. Amerasinghe, M. Chen, and K. Tantisirivat. Evaluating tight execution time bounds of programs by annotations. In *Proc. of the 6th IEEE RTOSS*, pages 74–80, May 1989.

87. Jaroslav Nadrchal. Architectures of parallel computers. http://www.fzu.cz/activities/schools/epsschool13/presentations/-parallel_architecture.pdf. In *Summer School on Computer Techniques in Physics*, 2002.

88. Chang Yun Park. Predicting program execution times by analyzing static and dynamic program paths. *Real-Time Systems*, 5(1):31–62, 1993.

89. Chang Yun Park and Alan C. Shaw. Experiments with a program timing tool based on source-level timing schema. *IEEE Computer*, 24(5):48–57, 1991.

90. D. Patterson, T. Anderson, N. Cardwell, R. Fromm, K. Keeton, C. Kozyrakis, R. Thomas, and K. Yelick. A case for intelligent RAM: IRAM. *IEEE Micro*, 17(2), April 1997.

91. David A. Patterson, Garth A. Gibson, and Randy H. Katz. A case for redundant arrays of inexpensive disks (RAID). In *SIGMOD Conference*, pages 109–116, 1988.

92. PEARL, Process and Experiment Automation Realtime Language, http://www.irt.uni-hannover.de/pearl/pearl-gb.html, 2006.

93. ProfiBus. System technical description. PROFIBUS brochure – order-no. 4.002. Technical report, http://www.profibus.com/, 1999.

94. PTB. Time and Standard Frequency Station DCF77 (Germany). http://www.ee.udel.edu/~mills/ntp/dcf77.html. Technical report, Physikalisch-Technische Bundesanstalt (PTB) Lab 1.21, Braunschweig, February 1984.

95. Peter P. Puschner and Christian Koza. Calculating the maximum execution time of real-time programs. *Real-Time Systems*, 1(2):159–176, 1989.

96. Real-Time for Java Expert Group. *Real-Time Specification for Java, 2nd Edition.* http://www.rtsj.org/specjavadoc/book_index.html, March 2005.

97. H. Rzehak. Real-time operating systems: Can theoretical solutions match with practical needs. In W. A. Halang and A. D. Stoyenko, editors, *Real-Time Computing*, pages 47–63. Springer, Berlin, Heidelberg, 1994.

98. R. Schild and H. Lienhard. Real-time programming in PORTAL. *ACM Sigplan Notices*, 15(4):79–92, 1980.

99. Jules I. Schwartz. The development of JOVIAL. *ACM SIGPLAN Notices*, 13(8), August 1978.

100. Lui Sha, Ragunathan Rajkumar, and John P. Lehoczky. Priority inheritance protocols: An approach to real-time synchronization. *IEEE Trans. Computers*, 39(9):1175–1185, 1990.

101. M. Skambraks. *A Safety-Licensable PES Architecture for Task-Oriented Real-Time Execution without Asynchronous Interrupts.* VDI Verlag GmbH, Düsseldorf, 2006.

102. D. J. Smith and K. G. Simpson. *Functional Safety.* Butterworth-Heinemann, Oxford, 2001.

103. Jack Stankovic and Krithi Ramamritham. The SPRING kernel: A new paradigm for real-time systems. *IEEE Software*, 8(3):62–72, 1991.

104. John A. Stankovic. Misconceptions about real-time computing. *IEEE Computer*, 21(10):10–19, October 1988.

105. John A. Stankovic and Krithi Ramamritham. Editorial: What is predictability for real-time systems. *Real-Time Systems*, 2(4):246–254, November 1990.

106. John A. Stankovic, Marco Spuri, Krithi Ramamritham, and Giorgo C. Buttazzo. *Deadline Scheduling for Real-Time Systems.* Kluwer Academic Publishers, 1998.

107. Neil Storey. *Safety Critical Computer Systems.* Addison Wesley, 1996.

108. Alexander Stoyenko. *A Real-Time Language With A Schedulability Analyzer.* PhD thesis, University of Toronto, December 1987.

109. Alexander D. Stoyenko and Wolfgang A. Halang. Extending PEARL for industrial real-time applications. *IEEE Software*, 10(4), 1993.

110. TNI. ControlBuild, PLC programming and virtual commissionning desktop for control engineers, 2006.

111. James E. Tomayko. *Computers in Space: Journeys With Nasa.* Alpha Books, March 1994.

112. Wilfredo Torres-Pomales. Software fault tolerance: A tutorial. Report, NASA Langley Research Center, 2000.

113. Domen Verber. *Object Orientation in Hard Real-Time System Development.* PhD thesis, University of Maribor, Slovenia, 1999.

114. Domen Verber. *Object Orientation in Hard Real-Time Systems Development. In Slovene, with extended abstract in English.* PhD thesis, University of Maribor, October 1999.

115. Domen Verber and Matej Šprogar. Generation of optimal timetables for time-triggered CAN communication protocol. In *Proceedings of WRTP2004*, Istanbul, 2004. IFAC, Elsevier.

116. Niklaus Wirth. *Programming in Modula-2, 3rd ed.* Springer Verlag, Berlin, 1985.

Index

Abbreviations

2oo3: two-out-of-three, triple modular redundancy

ALU: Arithmetic/Logic Unit

APU: Application Processing Unit

ASIC: Application–Specific Integrated Circuits

BDM: Background Debugging Mode

CAN: Controller Area Network

DCF77: Longwave time signal and standard-frequency radio station (D:Deutschland, C:long wawe, F: Frankfurt, 77: 77.5 kHz)

DMA: Direct Memory Access

DSP: Digital Signal Processor

EDF: Earliest Deadline First Scheduling Algorithm

EDM: Event-controlled Data Modifications

ETA: Event Tree Analysis

FB: Function Block

FDI: Fault Detection and Isolation

FMEA: Failure Modes and Effects Analysis

FPGA: Field–Programmable Gate Array

FTA: Fault Tree Analysis

FTC: Fault-Tolerant Cell

G-MRMC: Global Monitoring, Reconfiguration and Mode Control

HAL: High Level Assembly Language

HAZOP: Hazard and Operability Studies

HDL: Hardware Description Languages

HTML: Hypertext Mark-up Language

I/O: Input / Output (devices)

IEC: International Electrotechnical Commission

IFATIS: Intelligent Fault Tolerant Control in Integrated Systems; EU 5th FW research project

IRAM: Intelligent RAM

ISR: Interrupt Service Routine

JDN: Julian Day Number

JTAG: Joint Test Action Group (IEEE 1149.1: Standard Test Access Port and Boundary-Scan Architecture

LLF: Least Laxity First scheduling algorithm

M_PEARL: Mehrrechner PEARL (PEARL for Distributed Systems)

MC: Memory Cell

MIPS: Mega Instructions Per Second

MISRA: Motor Industry Software Reliability Association

MRMC: Monitoring, Reconfiguration and Mode Control

NMR: N-Modular Redundancy

NTU: Network Time Unit

OO: Object-Oriented

PCP: Priority Ceiling Protocol

Other titles published in this series (continued):

Printing: Krips bv, Meppel, The Netherlands
Binding: Stürtz, Würzburg, Germany